3-D VISUALIZATION FOR ENGINEERING GRAPHICS

Sheryl A. Sorby
Michigan Technological University

Kim J. Manner
University of Wisconsin–Madison

Beverly J. Baartmans
Michigan Technological University

Prentice Hall

PRENTICE HALL, Upper Saddle River, New Jersey 07458

Library of Congress Cataloging-in-Publication Data

Sorby, Sheryl Ann,
   3-D visualization for engineering graphics / Sheryl A. Sorby, Kim
J. Manner, Beverly Gimmestad Baartmans.
      p.   cm.
   Includes index.
   ISBN 0-13-191602-5
   1. Engineering graphics.   2. Visualization.   3. Three-dimensional
display systems.   I. Manner, Kim J.   II. Baartmans, Beverly
Gimmestad,.
   T357.S67   1998
   604.2'0285'6693—dc21   97-30396
                                                       CIP

Acquisition Editor: *Eric Svendsen*
Editorial/Production Supervision: *Barbara Kraemer*
Editor-in-Chief: *Marcia Horton*
Assistant Vice President of Production and Manufacturing: *David W. Riccardi*
Managing Editor: *Bayani Mendoza de Leon*
Manufacturing Buyer: *Donna Sullivan*
Manufacturing Manager: *Trudy Pisciotti*
Creative Director: *Paula Maylahn*
Art Director: *Jayne Conte*
Cover Designer: *Pat Woscyzk*
Editorial Assistant: *Andrea Au*
Compositor: *Preparé Inc. / Emilcomp srl*

© 1998 by Prentice Hall, Inc.
Upper Saddle River, NJ 07458

Printed in the United States of America

10   9   8   7   6   5   4

ISBN 0-13-191602-5

Prentice-Hall International (UK) Limited, *London*
Prentice-Hall of Australia Pty. Limited, *Sydney*
Prentice-Hall of Canada, Inc., *Toronto*
Prentice-Hall Hispanoamericana, S. A., *Mexico*
Prentice-Hall of India Private Limited, *New Delhi*
Prentice-Hall of Japan, Inc., *Tokyo*
Prentice-Hall Asia Pte. Ltd., *Singapore*
Editora Prentice-Hall do Brasil, Ltda., *Rio de Janeiro*

Dedicated to my children Katy, Lea, and Jordan.
I hope that someday you are as proud of me as I am of you.
S.A.S

Dedicated to my father Ernest, whose influence led to my interest in engineering.
K.J.M

Dedicated to my daughters, Maryann Gimmestad and Katherine Gimmestad.
B.J.B.

Acknowledgments

· ·

We would like to thank the National Science Foundation for its financial support of this project and the Michigan Tech Faculty Development program. We would also like to thank the following people who assisted us in the preparation of many of the figures for this text: Kim Ullman, Virgil Schlorke, Keri Ellis, Matt Buettner, Brian Garcia, Evan Berglund, Jason Geary, and Robert Ott. We would also like to thank Linda Ratts Engleman, formerly of Prentice Hall, who believed in this project and Eric Svendsen and Kate Cassino of Prentice Hall for their help in the completion of this work. Finally, we would like to thank our families and friends for their support during the many long hours of preparation of this text.

Contents

• • • • • • • • • • • • • • • • • • • •

PREFACE   *ix*

1. *ENGINEERING DESIGN GRAPHICS*   **1**

2. *POINTS, LINES, AND PLANES IN SPACE*   **6**

     2.1 3-D Coordinate Systems   6
     2.2 Points in 3-D Space   9
     2.3 Sketching Lines in 3-D Space   11
     2.4 Sketching Planes in 3-D Space   13
     2.5 Applications   15

3. *PICTORIAL SKETCHING*   **25**

     3.1 Technical Sketching   25
     3.2 Construction and Isometric Sketching of Buildings   27
     3.3 Oblique Sketching   33
     3.4 Lettering   37
     3.5 Applications   39

4. *TRANSFORMATION OF 3-D OBJECTS*   **44**

     4.1 Translation of Objects   44
     4.2 Dilation of Objects   48
     4.3 Rotation of Objects About a Single Axis   49
     4.4 Rotation of Objects About Two or More Axes   54

4.5 Reflection of Objects   60
4.6 Application   64

5. ORTHOGRAPHIC PROJECTION   **70**

5.1 Normal Surfaces   72
5.2 Inclined Surfaces   88
5.3 Oblique Surfaces   98
5.4 Single-Curved Surfaces   108

6. DRAWING STANDARDS AND CONVENTIONS   **117**

6.1 Scales   117
6.2 Conventional Practice and Orthographic Projection   125
6.3 View Selection   129

7. AUXILIARY VIEWS   **134**

7.1 Creating Auxiliary Views   134
7.2 Auxiliary Views of Curved or Irregular Surfaces   138
7.3 Applications   142

8. SECTION VIEWS   **166**

8.1 Section Types   169
8.2 Rib Features in Section   183
8.3 Application   185

9. DIMENSIONING   **191**

9.1 Format of Dimensions   191
9.2 Dimension Placement   202
9.3 Dimensioning Technique   209
9.4 Tolerances   214

10. WORKING DRAWINGS   **220**

10.1 Reading Construction Drawings   220
10.2 Mechanical Engineering Working Drawings   240

INDEX   **253**

Preface

· ·

For several years, engineering educators have called for a 3-D solids approach to graphics education. Unfortunately, there have been few texts introduced which are based on a 3-D solid modeling approach to design and design representation. In most of the graphics books which currently exist, a traditional approach is taken. This approach has remained virtually unchanged for the past fifty or more years. These traditional texts are based on a design methodology which revolved around a drawing board and T-square. Recent attempts to add computer applications in many texts have focused on using the computer as an electronic drawing board. The traditional texts typically start out with orthographic projection and move on to pictorials, sections and conventions, dimensioning, and working drawings. This text intends to fill this gap in engineering graphics education. The focus of the first part of the text is on the development of the visualization skills which are necessary to effectively use solid modeling software. The latter half of the text deals with understanding engineering drawings where visualization is again the focus.

When we sat down to write this text, we looked at two things: 1) how traditional engineering graphics courses have been taught; and 2) how to change graphics education so that it is based on 3-D solids. With traditional graphics courses, the first topic which is covered is typically orthographic projection. From there, students move on to topics such as sections and conventions, thread specifications, working drawings, and so on. We asked ourselves: If you are teaching a solid-modeling-based graphics course, what is the first thing that students need to learn? The answer was: how points, lines, and planes are defined in space. Thus, the first chapter in this text focuses on points in space, but not from a descriptive geometry standpoint. Points in space are defined according to a cartesian coordinate system as would be expected in solid modeling software.

We wanted the students to be able to sketch 3-D objects, since if they are constructing a computer solid they would likely need to make sketches to illustrate their procedure for combining solids, extruding, and so on. Correspondingly, the next chapter in the text is on pictorial sketching. This is taught from the standpoint that students have a set of building blocks which they can use to create and then sketch objects. Although this works best if the students actually have the blocks to create the solids, many of them can work the problems without the use of the blocks. The blocks which we use are 2 cm Snap Cubes, which can be purchased by Cuisenaire Company or by any other supplier of mathematical educational manipulatives.

Additional topics that students need to understand if they are to use solid modeling software are object transformations (scale, translate, rotate and reflect). These are all covered in Chapter 4. At this point, students should have acquired all of the visualization skills necessary for the creation of 3-D solids.

The next step in the design process is to generate a drawing from a 3-D solid. At this point we introduce orthographic projection (Chapter 5) and some of the more "traditional" graphics topics such as conventional practice (Chapter 6), sections (Chapter 8), and dimensioning (Chapter 9). Chapter 10 focuses on reading and interpreting working drawings with both civil and mechanical emphases. This chapter is included to provide a basis for understanding how the existing 3-D computer generated geometry can be queried to obtain useful information.

We purposely left out all reference to specific software packages within this text. Faculty will have a choice of three solid modeling texts (with I-DEAS, Pro/ENGINEER, and AutoCAD Designer) which could be used in conjunction with this graphics text or as stand-alone texts. Many of the exercises throughout the text focus on visualization of objects rather than on the techniques used in creating drawings. We believe that the graphics education of the future will require a greater emphasis on the development of visualization skills, which is what we have tried to accomplish with this text.

We are sincerely grateful to the reviewers who have made valuable contributions to this project. The reviewers include: Clarence Teske of Virginia Polytechnic Institute; Scott Tolbert of University of North Dakota; Vera Anand of Clemson; and Julia Jones of the University of Washington.

Engineering Design Graphics

Engineers probably affect your life more profoundly than any other group of people. Engineers are responsible for the design and manufacture or construction of products which you use in nearly every aspect of your everyday life. Everyone knows that engineers design cars, airplanes and bridges, but few people realize the extent of the impact that engineers have on our lives. For example, consider a simple process such as brushing your teeth. An engineer designed the toothbrush which you used, and a materials engineer developed the materials that were used in its handle as well as its bristles. Another engineer designed the process that was used to produce the toothpaste with the correct ingredients and consistency. Engineers developed the processes that were followed to manufacture the toothbrush and the toothpaste tube, and other manufacturing engineers were responsible for figuring out how to fill the toothpaste tube and put the cap on. In addition, engineers designed the system which brought the fresh, clean water that you used to brush your teeth. Engineers designed the pumps and fixtures used in your home water system that allowed the water to flow out of the faucet when you turned the handle. When you finished brushing your teeth, wastewater flowed down the drain and was collected and treated in a sanitation system that was also designed by engineers. The toothbrush and toothpaste were brought to your supermarket in vehicles designed by engineers over a transportation system that was also designed by engineers.

One hundred years ago, if you were a farmer in need of a plow blade, you went to the local blacksmith and described the type of blade you needed. The blacksmith, as an artisan, would then imagine what the blade looked like and make it. Engineers typically are not artisans. They will imagine a solution to a problem, but they usually will not manufacture the solution. For example, a mechanical engineer may have a specific gear arrangement in mind in the design of an engine. However, he or she will probably not be the person who creates the necessary parts for the design solution. Instead, the engineer will hand over the completed design to a machinist who will make the parts. In order for the machinist to create the parts which the engineer has imagined, there must be communication of the design idea between the two.

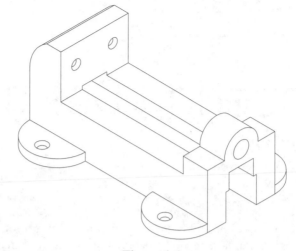

Figure 1-1

In the engineering profession, graphical communication is perhaps the most important form of relaying ideas from one person to another. In engineering, a picture truly is worth a thousand words. One picture can easily convey a design idea that might take several paragraphs or pages to describe with words. For example, Figure 1-1 shows an object to be manufactured. Try to imagine how many words would be necessary to describe this object if pictures were not allowed. Now try to imagine how many words would be necessary to convey the design ideas in a complicated structure such as the Mackinac Bridge shown in Figure 1-2.

Figure 1-2 Courtesy of Mackinac Bridge Authority

If engineers used words alone to describe their designs, imagine how confusing it would be for the craftsmen responsible who were machining or constructing them. There would be a lot of room for misinterpretation and errors.

There are three types of drawings that you may encounter as a practicing engineer: 1) schematic drawings, 2) line drawings, and 3) engineering drawings or blueprints. Schematic drawings do not usually look like the system they represent. They may show the flow of information or material, and schematic drawings often have standard symbols which are used to represent real-life objects. Since schematic drawings do not look like the objects they represent, they are usually not drawn to scale, but they are made to look proportionally correct. The simple electrical circuit diagram in Figure 1-3 shows an example of a schematic drawing.

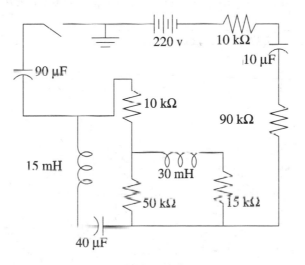

Figure 1-3

Line drawings are used to illustrate an idea or a system in a simplified manner. Line drawings are frequently seen as figures in papers or books. They usually do not include any dimensions, but may include one or two if they are important. You will usually try to make a line drawing which appears similar to the object or system you are trying to describe, but it won't be identical. Figure 1-4 shows a line drawing of a lathe with all of the major features labeled. If you are required to write a paper or a report, you will most likely need to

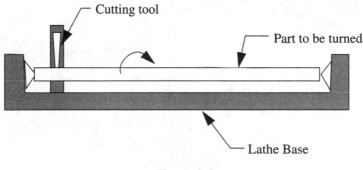

Figure 1-4

include one or more line drawings. Fortunately, with most word processors today, you will have the ability to make simple line drawings directly within your document.

Technical drawings or blueprints are complex drawings which contain dimensions, labels, and notes in addition to displaying the basic part geometry. The object or system to be produced is usually shown from many different vantage points. Engineering drawings typically look like the object they are trying to portray and contain all or nearly all of the information required to produce the part. Figure 1-5 shows a drawing of a Brake Bridge from a mountain bike. Notice that this drawing shows the part from several different points of view—the front, the side, and the back. A pictorial wireframe of the part is also included in the upper right corner of the drawing for clarity. Many times, a 3-D non-transparent view of the object would be shown instead of a wireframe. Dimensions and sizes are also given on this drawing so that it can be produced to exact specifications.

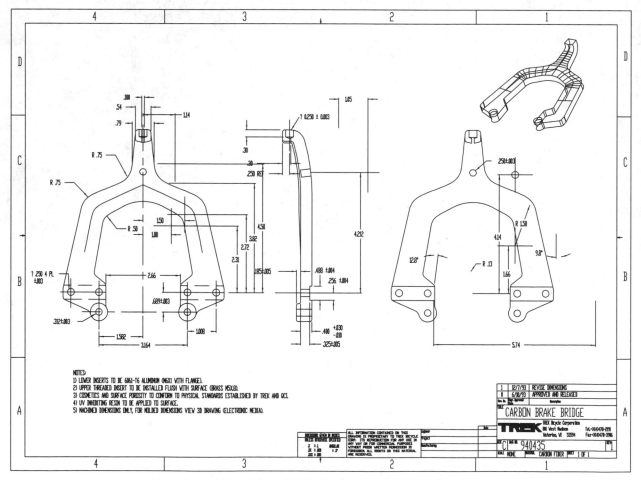

Figure 1-5 Courtesy of TREK Bicycle Corp

Engineers use graphical methods of communication to convey their design ideas, to avoid confusion, and to provide manufacturing or construction details. In many cases, the final and preliminary drawings are considered legal documents. They convey a great deal of information and can be used to establish patent and trade secret ownership. Drawings will also be used in the case of litigation related to the manufacture or construction of engineered systems. Since graphical communication is such an important part of engineering practice, it is important for you to learn about the standards and practices involved in this type of communication. This is similar to the way that you learned spelling and grammar rules in order to convey your ideas effectively in written communication. This text will introduce you to the field of Engineering Design Graphics. Although you may not be responsible for *creating* complicated engineering drawings as a practicing engineer, you will be responsible for checking drawings to make sure that they accurately portray your design intent. You will also need to know how to make sketches in the preliminary stages of the design process and how to read and understand drawings made by others.

Earlier graphics texts were often focused on the "how's" of graphical communication—how to draw a line tangent to a circle, how to draw an ellipse by constructing four circular arcs, how to draw the different types of section lining, and so on. The emphasis of this text will be on the "why's" of graphical communication—why sectional and detail views are included in a set of plans, why dimensions are placed in one view instead of another, why certain views are necessary, and so forth.

Drawing with instruments, either a T-square and pencil or a computer and mouse, will not be described in this text. The mechanics of *how* complex technical drawings are made will not be presented here. Instead, sketching will be stressed almost exclusively and the understanding of engineering drawings and conventions is the primary focus of this text.

C H A P T E R 2

Points, Lines, and Planes in Space

• •

Engineering drawings are made to represent three-dimensional objects. In order for you to better understand how to read drawings, it is important to first look at basic three-dimensional geometric entities in space. This chapter will present points, lines, and planes in space and show how these three-dimensional entities can be represented on a two-dimensional sheet of paper.

2.1 3-D COORDINATE SYSTEMS

Before looking at points, lines, and planes in space, it is necessary for you to understand some of the conventions used to portray 3-dimensional space on a 2-dimensional sheet of paper. The *isometric* coordinate axes and the *oblique* coordinate axes are two different representations of coordinate systems commonly used in portraying 3-dimensional space on paper. In 3-dimensions, the X-axis, the Y-axis, and the Z-axis are all mutually perpendicular. However, the relationship of these axes when projected onto a 2-D surface, such as a piece of paper, is portrayed differently using isometric and oblique representations. Figure 2-1 shows the orientation of the coordinate axes in each of these systems.

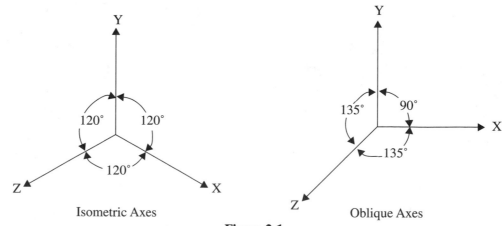

Isometric Axes   Oblique Axes

Figure 2-1

It is important to be able to visualize the axes of the XYZ coordinate system as being mutually perpendicular in 3-dimensions. With isometric axes, think of the X- and Z-axes as perpendicular to one another and lying together in a plane with the Y-axis perpendicular to that plane. With oblique axes, visualize the X- and Y-axes as perpendicular to one another lying in a plane with the Z-axis perpendicular to that plane. Figure 2-2 shows these two interpretations.

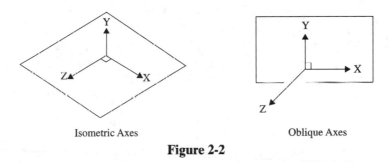

Isometric Axes   Oblique Axes

Figure 2-2

The actual labeling of axes may also differ depending on academic discipline and topic of discussion. Figure 2-3 shows two common labeling schemes of the X-, Y-, and Z-axes for two sets of oblique coordinate axes.

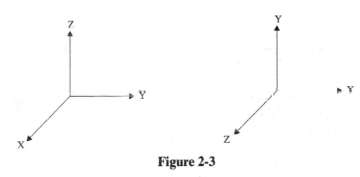

Figure 2-3

All of the coordinate systems discussed so far have axes that follow the *right-hand rule*: If you place the fingers of your right hand along the positive X-axis and curl them in the direction of the positive Y-axis, then the thumb of your right hand will point in the direction of the positive Z-axis (see Figure 2-4).

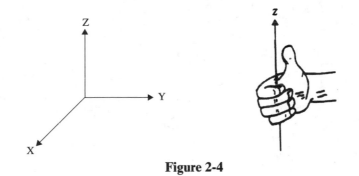

Figure 2-4

You will occasionally see a coordinate system labeled using the *left-hand rule*: If you place the fingers of your *left* hand along the positive X-axis and rotate them in the direction of the positive Y-axis, then the thumb of your left hand will point in the direction of the positive Z-axis. Figure 2-5 shows isometric coordinate axes labeled using the left-hand rule. In the next sections, you will see both isometric and oblique coordinate axes being used, but all labeling will be done using the right-hand rule.

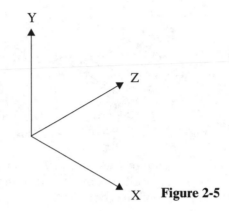

Figure 2-5

Isometric grid paper is a popular tool in engineering drawing for portraying 3-D space. Figure 2-6 shows two different types of isometric grid paper. In the first type of isometric grid paper, grid lines are evenly spaced on the page. The lines are parallel to the X, Y, and Z axes as displayed in isometric form. In the second type of grid paper, only dots are shown arranged in a pattern such that each dot represents the intersection of evenly spaced lines parallel to the coordinate axes. The bold axes shown in Figure 2-6 are for illustration purposes only.

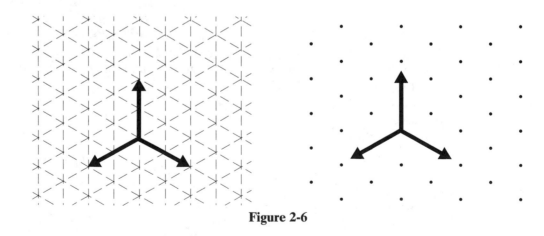

Figure 2-6

EXERCISES 2.1

In Exercises 1 through 4, identify each of the following coordinate axes as isometric or oblique and as right-handed or left-handed.

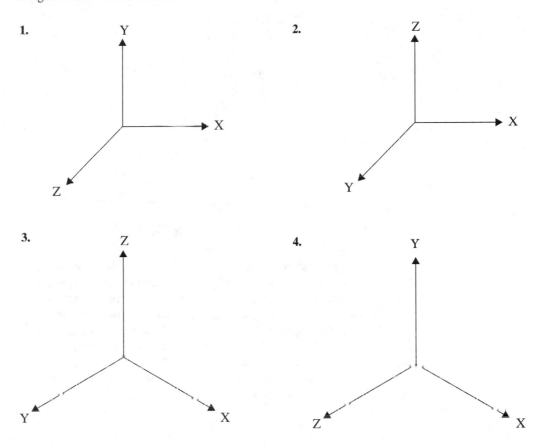

5. In Exercise 1, which axis should be visualized as being perpendicular to the plane of this page?
6. In Exercise 3, which axis should be visualized as being in the plane of this page?

2.2 POINTS IN 3-D SPACE

Points in 3-D space require three coordinates to define their locations—an X-coordinate, a Y-coordinate, and a Z-coordinate. Plotting a three-coordinate point, P(x, y, z), is similar to plotting a two-coordinate point, P(x, y), except that, in addition to moving in the X- and Y-directions, you must also move in a third direction—the Z-direction. Figure 2-7 shows several points plotted in 3-D space.

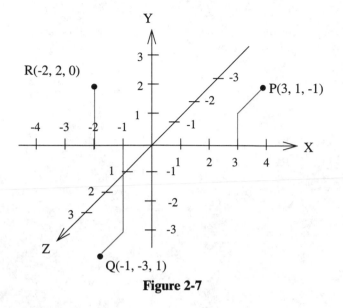

Figure 2-7

When you use isometric grid paper to represent 3-D points you have to be careful, because in this case, appearances can be deceiving. Because isometric grid paper is a 2-D representation of 3-D space, each dot on the grid paper represents more than one location in space. For example, the point shown in Figure 2-8 can have several possible 3-D coordinates that appear to describe it. Possible coordinates for this point include: $(-2, 0, 0)$, $(0, 2, 2)$, $(1, 3, 3)$, or $(2, 4, 4)$. In reality, with isometric grid paper, you are looking down a line in space and "seeing" the end view of the line. This line is oriented in space so that it corresponds to the diagonal of a cube. Thus, each grid point represents an end view of a line in space rather than a single point.

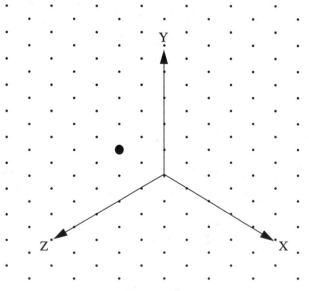

Figure 2-8

EXERCISES 2.2

1. Plot the following points on an isometric axis. Label each point.

A(3, 1, 4)   B(2, 1, −3)   C(1, −2, 0)   D(−4, 1, −2)
E(0, 2, −1)   F(2, −3, 1)   G(3, −2, −1)   H(−4, −2, −1)
I(2, 0, 1)

2. For each point in the previous exercise, list two other possible points which "appear" to be at that location in space.

2.3   SKETCHING LINES IN 3-D SPACE

Lines are one-dimensional geometric entities in space. Two points are necessary to define a line, and lines are theoretically infinite in length. In this text, we will represent lines as finite segments defined by their two endpoints. Lines are drawn in 3-D space by first locating their two endpoints and then drawing a line between the two points. Figure 2-9 shows two lines in 3-D space. Line AB is drawn between points A(0, 2, 1) and B(3, 1, −1), and line CD was drawn between points C(−2, 1, 2) and D(2, −2, 0).

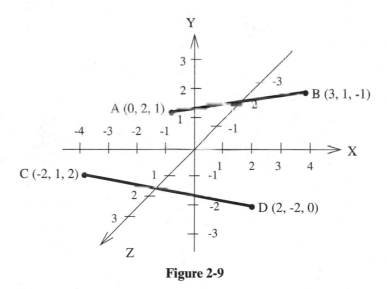

Figure 2-9

When sketching lines which seem to intersect but are actually located on different planes, it is customary to draw the line "in front" as a continuous line and to make a break in the line which is "in back." This problem occurs because we are representing 3-D space on a 2-D sheet of paper. Otherwise the false impression is given that the lines actually lie in the same plane and are intersecting lines. Figure 2-10 shows two lines, PQ and RS. The coordinates which define the points are shown in the figure. Notice that the line PQ is shown with a gap around it in the vicinity of line RS. This is due to the fact that line PQ lies behind RS. To draw a solid line at the implied intersection would signify that these lines lie in the same plane and not one behind the other, as shown in this figure.

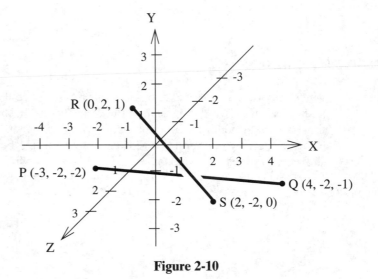

Figure 2-10

EXERCISES 2.3

1. Examine the graphs shown below and answer the following questions.

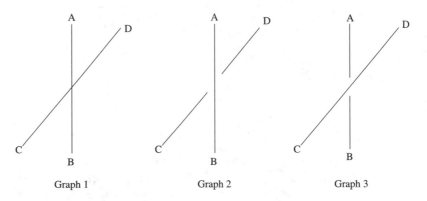

Graph 1   Graph 2   Graph 3

   a. In which graph does AB lie behind CD ?
   b. In which graph does AB cross CD ?
   c. In which graph does CD lie behind AB ?

2. Sketch the following lines on isometric grid paper. Use the point locations as specified in problem #1 of Exercises 2.2.

   AB   CF   DG   EI   HA

2.4 SKETCHING PLANES IN 3-D SPACE

The easiest planes to name and graph in 3-D space are the coordinate planes and planes parallel to them. Figure 2-11 shows the three coordinate planes and their respective equations.

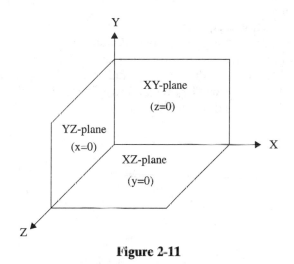

Figure 2-11

In two dimensions, an equation like $y = 2$ means that the y-coordinate is always two, but the x-coordinate has no constraint on it or is said to be "free to vary." Hence, the equation $y = 2$ in two dimensions is a line. In particular, it is a line parallel to the X axis but elevated two units in the positive Y-direction. In three dimensions, the equation $y = 2$ means that the y-coordinate is always two, but the x-coordinate and the z-coordinate are both free to vary. Hence, in three dimensions, the equation $y = 2$ is a plane parallel to the XZ-plane but elevated two units in the positive Y-direction. Figure 2-12 shows the graph of the plane $y = 2$ as well as the graphs for the planes $x = -4$ and $z = 3$. These planes, although they appear finite in size, are theoretically infinite.

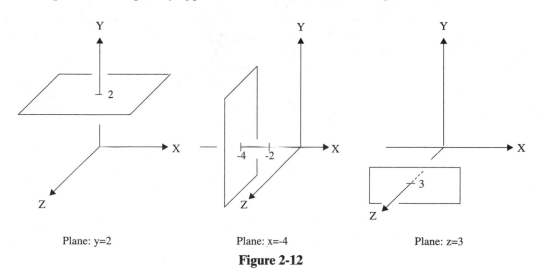

Plane: y=2   Plane: x=-4   Plane: z=3

Figure 2-12

There are some rules about sketching planes that are important to follow for effective visual communication. Notice that when you draw planes parallel to the coordinate planes, they are drawn as parallelograms with sides parallel to the coordinate axes. Further notice that portions of the coordinate axes hidden by the planes either are not drawn (as in the graphs of y = 2 and x = −4) or are drawn using dotted lines (as in the graph of z = 3).

Some equations for planes have only two variables. These equations look like the equations for lines in two dimensions, but because they are being graphed in three dimensions, they are actually planes. The missing third variable is free to vary, so the line slides in the direction of that variable, creating a plane. For example, the plane 3x + 2y = 6 can be sketched as a line in the XY-plane. Sliding the line along the Z-direction creates the graph of the plane (see Figure 2-13).

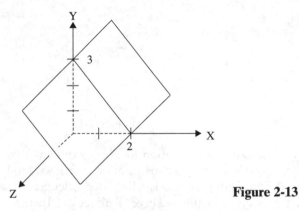

Figure 2-13

Planes that are not coordinate planes (or parallel to coordinate planes) will eventually intersect two or all three of the coordinate axes. The general equation for a plane is ax + by + cz = d. The points where a plane intersect the X-, Y-, and Z-axes easily determined. Along the X-axis, y and z are both 0, so the point where the plane crosses the X-axis will occur at the point on the plane where y and z are both set to 0. You can use a similar procedure to determine the points where the plane crosses the Y- and Z-axes. The plane is then sketched by connecting the three points with lines. Figure 2-14 shows the graph of the plane 2x + 3y + 4z = 12.

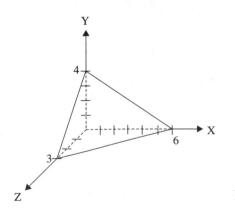

Figure 2-14

EXERCISES 2.4

1. Graph the planes $z = 4$ and $z = -4$ on isometric grid paper.
2. Graph the plane $2x - y + 5z = 10$.
3. Graph the plane $3y + 5z = 15$. Hint: $3y + 5z = 15$ can be graphed as a line in the YZ-plane, which slides in the X-direction to create a plane because x is free to vary.

2.5 APPLICATIONS

Land Surveying. Surveying has long been used in measuring and marking land for personal ownership. Ancient Babylonians and Egyptians used crude measurement devices to establish landmarks and distinguish different people's property. Nineteenth-century pioneers in the opening West measured their homesteads by pacing off the specified distances. Surveying has evolved into an exacting science with precision instruments capable of measuring distances to the nearest thousandth of a foot and measuring angles to fractions of seconds. Unlike the homesteads, which were typically square or rectangular parcels of land, most property is not regularly shaped. When mapping out a parcel of land, a surveyor constructs what is known as a **traverse**. In surveying terms, a traverse is a multi-sided area of land that starts and stops at the same point. Thus, a traverse is a plane in space. When taking field data for a traverse, the surveyor determines both the distance between the points that define the traverse and the bearing (defined in the following) of each of the edges of the traverse. This information is then used in the computation of the area of the traverse. The mathematical technique used in calculating the area of the traverse is beyond the scope of this text, but sketching the traverse itself is possible using the techniques previously described in this chapter.

In analyzing the field data to complete an accurate drawing of the traverse, a **bearing** of a line is defined as the angle the line makes with a true north–south line. The bearing of a line also indicates its direction with the line having a starting point and an ending point. Figure 2-15 shows how bearings of lines are defined. Note that the bearing of line

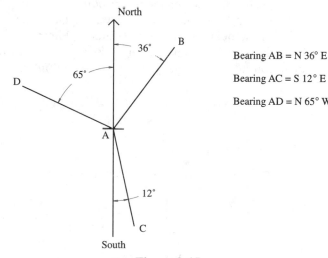

Figure 2-15

AB is N 36° E, but that the bearing of line BA is S 36° W. Similarly, the bearing of line CA is N 12° W and the bearing of line DA is S 65° E.

Figure 2-16 shows a five-sided traverse, and Table 2-1 gives the field data obtained by the surveyor. If A is considered to be the origin of the coordinate system, going from point A to the opposite endpoint of any given line changes both x and y. Changes in the north and east directions are analogous to positive changes of x and y; changes in the south and west directions are analogous to negative changes of x and y. The amount of change in x and y between point A and point B is related to the length and the bearing of the line AB.

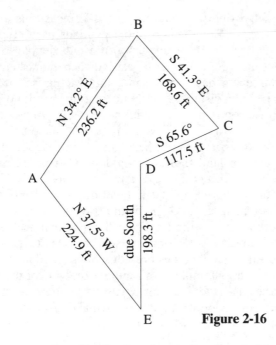

Figure 2-16

TABLE 2-1

Side	Length	Bearing
AB	236.2 ft.	N 34.2° E
BC	168.8 ft.	S 41.3° E
CD	117.5 ft.	S 65.6° W
DE	198.3 ft.	due South
EA	224.9 ft.	N 37.5° W

Specifically, the change in x is equal to the length of the line multiplied by the sine of the bearing, and the change in y is equal to the length of the line multiplied by the cosine of the bearing. This is illustrated in Figure 2-17. In surveying terminology, changes along the east–west line (changes in x) are called **departures** and changes along the north–south line (changes in y) are called **latitudes**.

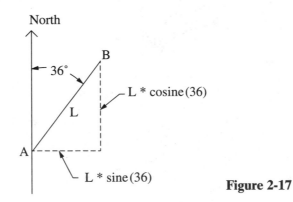

Figure 2-17

When compiling traverse data, it is important to note that the sum of all of the departures as well as the sum of all of the latitudes must equal zero because a traverse begins and ends at the same point. The departures and latitudes from traverse data can also be used to determine the X-, Y-locations of the points around the traverse. Table 2-2 shows the data obtained from the traverse in Figure 2-16.

TABLE 2-2

Line	Length	Bearing	Departure		Latitude	
			+ East	− West	+ North	− South
AB	236.2 ft	N 34.2° E	132.7		195.3	
BC	168.8 ft	S 41.3° E	111.3			126.8
CD	117.5 ft	S 65.6° W		107.1		48.5
DE	198.3 ft	due South				198.4
EA	224.9 ft	N 37.5° W		136.9	178.4	
		Sums	244.0	−244.0	373.7	373.7

Note that the sums of the latitudes and departures are indeed equal to zero, which indicates that the surveyor started and stopped at the same point. If point A is selected as the origin for the traverse, the X- and Y-coordinates of the subsequent points in the traverse equal the cumulative sum of the departures and latitudes, respectively, up to the point under consideration. Thus, for point C, the X-coordinate is equal to the departure for AB plus the departure for BC, or 132.7 + 111.3 = 244.0 (the Y-coordinate is equal to the sum of the latitudes, i.e., 195.3 − 126.8 = 68.5). Table 2-3 includes the X- and Y-coordinates for each of the points in the traverse. Note that the value of the Z-coordinates of the plane which contains the traverse are all zero.

TABLE 2-3

Point	X	Y
A	0.0	0.0
B	132.7	195.3
C	244.0	68.5
D	136.9	20.0
E	136.9	−178.4
A (check)	0.0	0.0

Electrical Conduits.   In designing a new building, an electrical engineer must see that power is distributed throughout the building in a logical manner. Typically, the power source is distributed vertically through the building and electrical conduits connect horizontally to the source on the floors of the building. Sometimes a piece of equipment on one of the floors has significant power needs. A separate conduit might be located on the floor just for that piece of equipment.

It is useful to be able to visualize how power is distributed throughout a building. You can then determine the shortest path for future wiring projects. You can also determine whether there are unnecessary cables in the building if any area of the building has too many conduits.

Because the power source is a vertical line through the building, it is useful to think of it as extending along the Y-axis. The power conduits that are connected to the source then extend in the X- and Z-directions. For example, consider a five-story building approximately 50 feet in height. If you plot the main power source for the building using isometric grid paper, and use a scale of one grid point equal to 10 feet, the drawing shown in Figure 2-18 results.

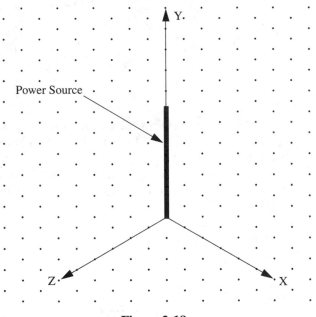

Figure 2-18

In this case, the conduit lines all begin at the power source and extend in the X- and Z-directions. Therefore, the starting point of each conduit line occurs at a given Y-value and the beginning X- and Z-values are both zero. The ending point for each conduit line is at the same Y-value as the starting point, but at different X- and Z-values. Each conduit that extends from the power source can then be plotted as a line in three-dimensional space. For example, Table 2-4 contains the locations of the conduit lines for this building. To plot the first conduit line (A), start at the Y-location on the power source and draw a line to the point $x = 0, y = 0, z = 30$. Recall that the distance between grid points is equal to 10 feet. When conduit A is included, the drawing should look like that shown in Figure 2-19.

TABLE 2-4

Conduit Line	Y-Location	End Point	
		X-Location	Z-Location
A	0'	0'	30'
B	10'	40'	0'
C	10'	20'	30'
D	20'	10'	20'
E	30'	30'	10'
F	40'	40'	0'
G	50'	40'	0'
H	50'	0'	30'

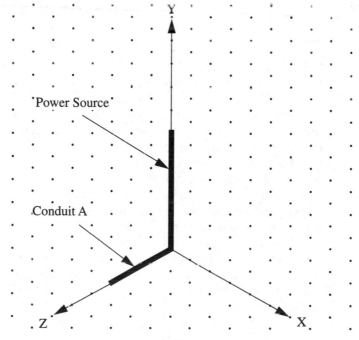

Figure 2-19

Figure 2-20 shows the electrical power supply diagram when all of the conduits have been included. Each conduit is labeled in the figure.

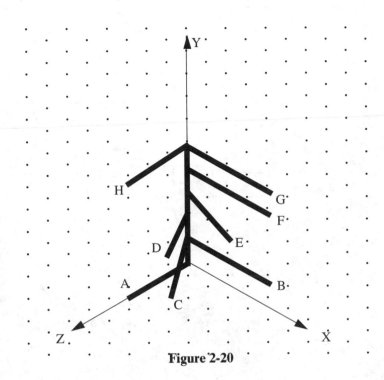

Figure 2-20

Wireframe Geometry.   Wireframe geometry is defined as 3-D geometry where the points defining the wireframe are connected by stiff wires similar to the frame for a box kite. Thus, the resulting 3-D geometry is hollow, not solid, and you are able to see through it. One way to draw a wireframe of a 3-D object is by plotting the corners of the object and then connecting pairs of corners with line segments. As you sketch the geometry, it is sometimes necessary to "move" from one point to another without drawing a line between the two. This situation is recorded in the Move/Draw column as a 0; whereas, drawing a line between the two points is recorded as a 1. In other words, a zero opposite a point means that you arrived at that point from the previous point by "picking up" the pen and moving it there, rather than by drawing a line from the previous point. Thus, the corners are points in three-dimensional space and a Move/Draw command tells which points to connect with line segments. A wireframe drawing of a 3-D object is shown in Figure 2-21. Vertices of the object can be assigned X-, Y-, and Z-coordinates. A partial listing of the vertices and their coordinates as you trace out the 3-D wireframe are shown in Table 2-5. In the exercises, you will be asked to fill in the missing entries.

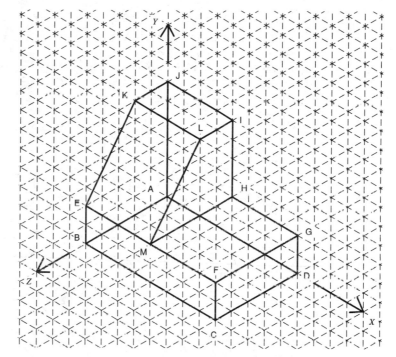

Figure 2-21

TABLE 2-5

Point	Move 0 Draw 1	X	Y	Z
A	0	0	0	0
B	1	0	0	5
C	1	8	0	5
D	1	8	0	0
A	1	0	0	0
B	0	0	0	5
E	1	0	2	5
F	1	8	2	5
G	1	?	?	?
H	1	4	2	0
I	1	4	6	0
J	1	0	6	0
K	1	0	6	2
L	1	?	?	?

Point	Move 0 Draw 1	X	Y	Z
I	1	4	6	0
L	0	?	?	?
M	1	4	2	5
K	0	0	6	2
E	1	0	2	5
J	0	0	6	0
A	1	0	0	0
M	0	4	2	5
H	1	4	2	0
F	0	8	2	5
C	1	8	0	5
G	?	?	?	?
D	?	8	0	0

EXERCISES 2.5

1. Determine the X- and Y-coordinates of each of the points A through E in the traverse shown below. For the latitudes and departures to sum exactly to zero, you may have to adjust some of the values (error is due to round-off in the data). Make all changes to line EA so that the departures and latitudes sum to zero.

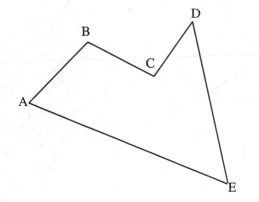

Line	Length	Bearing
AB	263.4 ft.	N 44.3° E
BC	227.9 ft.	S 62.0° E
CD	207.1 ft.	N 35.3° E
DE	518.4 ft.	S 11.9° E
EA	663.8 ft.	N 67.2° W

2. Determine the X- and Y-coordinates of each of the points A through F in the traverse shown below. You may have to adjust some of the latitudes and departures to make them sum to zero (error is due to round-off). Using small square grid paper, with one grid square equal to 20 feet, plot the traverse. Count the number of squares inside the traverse to estimate the traverse area.

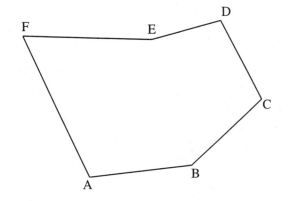

Line	Length	Bearing
AB	234.2 ft.	N 83.4° E
BC	223.0 ft.	N 46.8° E
CD	203.0 ft.	N 27.7° W
DE	164.0 ft.	S 74.9° W
EF	296.9 ft.	N 88.3° W
FA	360.0 ft.	S 25.3° E

3. Draw the power source and conduit lines for a building according to the following specifications. The building height is 40 feet. Use a grid spacing of one grid space equals 5 feet. The power source line extends along the Y-axis from the origin to a height of 40 feet. The conduit lines are defined in the following table. Label each conduit line in your drawing.

Conduit Line	Y-Location	End Point	
		X-Location	Z-Location
A	0′	20′	0′
B	10′	20′	15′
C	10′	25′	5′
D	20′	10′	0′
E	20′	30′	10′
F	30′	5′	25′
G	40′	10′	25′
H	40′	20′	0′

4. Draw the power source and conduit lines for a building according to the following specifications. The building height is 70 feet. Use a grid spacing of one grid space equals 10 feet. The power source line extends along the Y-axis from the origin to a height of 70 feet. The conduit lines are defined in the following table. Label each conduit line in your drawing.

Conduit Line	Y-Location	End Point	
		X-Location	Z-Location
A	0′	40′	50′
B	10′	30′	60′
C	20′	0′	50′
D	30′	20′	40′
E	40′	50′	0′
F	50′	50′	20′
G	60′	0′	40′
H	70′	30′	50′

5. Complete the missing entries in Table 2-5 for vertices G and L and the missing Move/Draw commands for the last two vertices of the table.

6. On isometric grid paper, draw the 3-D wireframe from the data given in the table below. Label each point.

Point	Move 0 Draw 1	X	Y	Z
A	0	0	0	0
B	1	0	3	0
C	1	0	4	4
D	1	0	1	4
E	1	1	0	3
F	1	2	0	0
A	1	0	0	0
D	1	0	1	4
C	0	0	4	4
E	1	1	0	3
B	0	0	3	0
F	1	2	0	0

7. On isometric grid paper, draw the 3-D wireframe from the data given in the table below. Label each point.

Point	Move 0 Draw 1	X	Y	Z
A	0	0	0	0
B	1	0	0	4
C	1	0	4	4
D	1	0	4	0
E	1	2	4	0
F	1	2	4	4
C	1	0	4	4
F	0	2	4	4
G	1	2	0	4
B	1	0	0	4
G	0	2	0	4
O	1	2	0	0
E	1	2	4	0
L	0	2	3	0

Point	Move 0 Draw 1	X	Y	Z
M	1	2	3	2
H	1	2	0	2
I	1	4	0	2
N	1	4	3	2
M	1	2	3	2
L	0	2	3	0
K	1	4	3	0
N	1	4	3	2
K	0	4	3	0
J	1	4	0	0
I	1	4	0	2
J	0	4	0	0
A	1	0	0	0
D	1	0	4	0

Pictorial Sketching

Pictorial sketching is used to portray an object as it appears in 3-D space. Pictorial sketches are useful for showing a 3-D representation of an object on a flat (2-D) piece of paper. There are two basic types of pictorial sketches which you will make as engineers: isometric and oblique. In this chapter you will learn how to make these types of sketches. We will focus on creating pictorial sketches for objects which can be constructed out of blocks. In later chapters of this text, you will learn about objects that contain slanted, or inclined, surfaces. At that time, you will be expected to be able to create pictorial sketches of these more complex objects. You will also learn the basics of lettering in this chapter, since this is an important skill for you to acquire as you prepare your sketches.

3.1 TECHNICAL SKETCHING

Many times in your engineering career you will be called upon to create sketches which illustrate your design. Most technical sketches are relatively simple to construct. You do not have to worry about exact dimensions the way that you do for a technical drawing, but you should draw carefully so that your sketch is clear, easy to read, and shows proper proportions of the object features. In order to make accurate sketches, there are a few rules of freehand sketching which you may find useful.

In drawing lines, the key is to have straight and not "wavy" lines. It is easier to draw straight lines if you draw vertical lines in a downward motion and if you draw horizontal lines to the right. If you are a left-hander, you may find it easier to draw straight horizontal lines going from right to left across the paper. To draw inclined lines, it is generally easier to sketch a straight line which is angled up as you go from left to right. For inclined lines which angle down as you go from left to right, your hand will probably be "in the way" as you sketch, and therefore, it will be more difficult to keep the line straight. For this type of line, you should probably turn the paper so that you are drawing in either a downward or a right-to-left motion. When you are sketching long lines, it is best if you keep your eyes focused on the endpoint to which you are sketching and to start out by lightly drawing the line. As the sketch takes form, you should darken in any lines which were lightly drawn. The techniques used for sketching straight lines are illustrated in Figure 3-1.

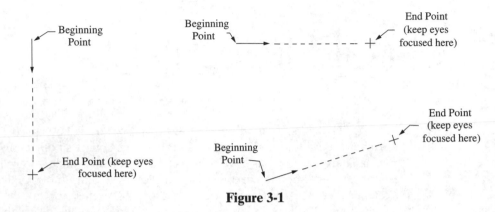

Figure 3-1

When making technical sketches, you will often need to sketch a circle or an ellipse as a part of the object you are trying to portray. With this type of curved entity, it is best if you first lightly sketch the bounding box for the curved entity. The bounding box should have the lengths of its sides equal to the diameter of the circle or ellipse. You then mark the midpoints of the sides of the bounding box and sketch in its arc segments tangent to the sides of this box and going through the midpoints which you marked. If the circle is particularly large, you may need to also add radial points at 45° angles (the radial points are at roughly 2/3 of the diagonal). For smaller circles, it may not be necessary to first mark the radial points. The technique for sketching circles and ellipses is illustrated in Figure 3-2.

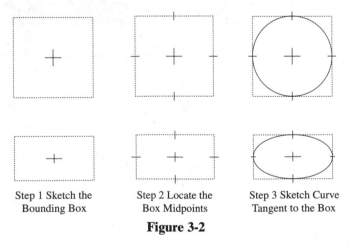

Step 1 Sketch the   Step 2 Locate the   Step 3 Sketch Curve
Bounding Box   Box Midpoints   Tangent to the Box

Figure 3-2

One of the most difficult tasks in creating good technical sketches is in making sure that the proportions of the features you are sketching are correct. This means that the overall height-to-width ratio of the sketch should look proportionally correct. Also, features which are small should remain small on the sketch relative to the larger features. If you do not keep the features of the object proportionally correct in your sketch, then the object will look distorted and your design ideas will not be properly conveyed. As with all of the sketching techniques described in this section, the key to making good, clear sketches is to practice. The more you practice your sketching technique, the better you will become. The remainder of this text will focus on visualizing and sketching objects from different vantage points. As you work the problems in this text, focus on good sketching technique. With practice, you will develop and improve your skills until good sketching technique becomes automatic.

EXERCISES 3.1

1. Freehand sketch a cube from several different viewpoints. Make sure that the features of the cube are proportionally correct in size.
2. Freehand sketch a cereal box from several different viewpoints. Make sure that the features of the cereal box are proportionally correct in size.
3. Freehand sketch a coffee cup from several different viewpoints. Make sure that the features of the coffee cup are proportionally correct in size.
4. Freehand sketch a table from several different viewpoints. Make sure that the features of the table are proportionally correct in size.

3.2 CONSTRUCTION AND ISOMETRIC SKETCHING OF BUILDINGS

To create isometric sketches, you might first want to construct the 3-D objects, observe these objects, and then draw them to scale from different viewpoints. As your spatial sense develops, it will no longer be necessary to construct an object before drawing it. A mental image of the object will be sufficient to create the sketch. Hence, you may want to initially work from the hand-held object, but ultimately, you will want to be able to visualize the object and create an accurate sketch of it from this mental image.

Cubes will be used in the following examples to construct model buildings. The plan for each building will be given by drawing the shape of the top (or base) of the building on square grid paper. Each square will be coded with a number to represent how high the stack of cubes on that square should be. For example, Figure 3-3 shows a coded plan and the corresponding building constructed from cubes as viewed from corner D. Note that this corner view of the object predominantly shows its front and left side.

A corner view of a building, such as the one shown in Figure 3-3, is called an **isometric** view of the building. Such views are easy to draw on isometric grid paper. With

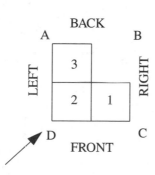

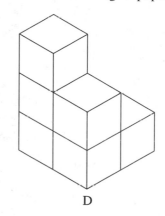

Figure 3-3

isometric projection, the viewing plane is located perpendicular to a diagonal of a cube and the corner points of the object are projected perpendicularly onto this plane. Figure 3-4 shows an isometric sketch of the object defined in Figure 3-3 drawn on isometric grid paper. Note that the isometric sketch has fewer lines than the object lines shown in Figure 3-3. When sketching an isometric view of a building, it is not necessary to outline each cube, rather a line is drawn only where there is an edge to the building. An **edge** can be defined as a line which results from the intersection of two plane surfaces.

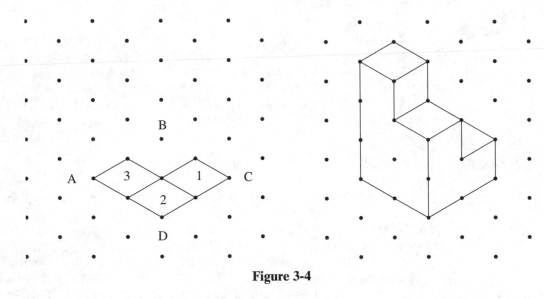

Figure 3-4

Sometimes it is easier to think of an object as a collection of surfaces. The isometric drawing of the object can then be made by sketching each visible surface. This process is demonstrated in Figure 3-5, where each surface is drawn, one at a time, until the sketch is complete. In step 1 the left side surface is sketched. In step 2, the foremost front surface is sketched. In steps 3 and 4 the uppermost top surface and the inside front surface are sketched. Finally, in steps 5 and 6, the two remaining top surfaces which define the object are sketched.

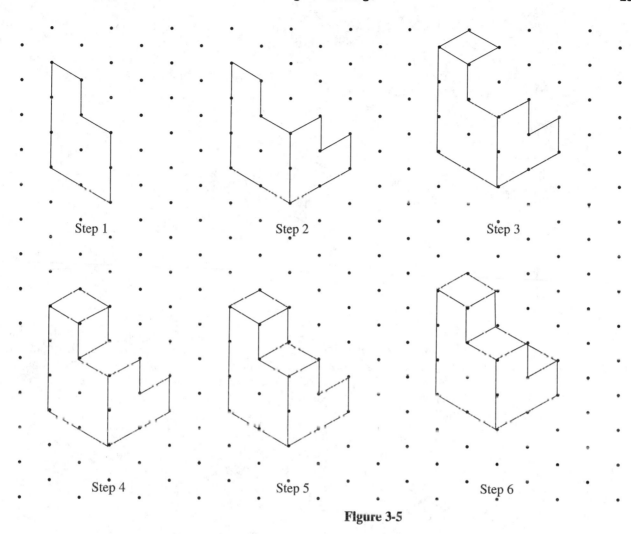

Step 1

Step 2

Step 3

Step 4

Step 5

Step 6

Figure 3-5

Figure 3-6 shows another coded plan, the corresponding building constructed from cubes, and the corresponding isometric sketch of the building. This time the building is being viewed from corner C. Note that this corner view shows the object's front and right side views. Typically, you will choose the appropriate isometric viewpoint which shows the object features more distinctly.

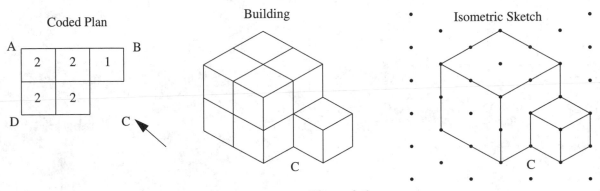

Figure 3-6

Isometric sketches of objects can be made using grid paper by using a thicker or softer pencil lead on the paper or by placing a piece of tracing paper over the isometric grid paper. Figure 3-7 shows the four corner views of a building on isometric grid paper. The coded plan for the building can be determined by viewing the building from its corners and is given in Figure 3-8. Note that the corner views "A" and "C" show the object more distinctly.

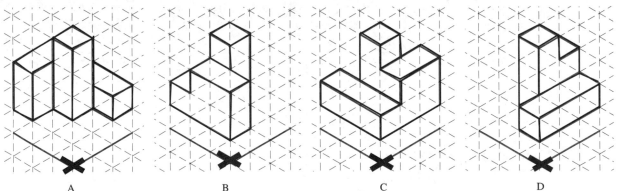

Figure 3-7

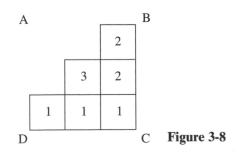

Figure 3-8

EXERCISES 3.2

In Exercises 1 and 2, sketch the coded plan for the building shown. Assume there are no hidden cubes.

1.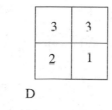

2.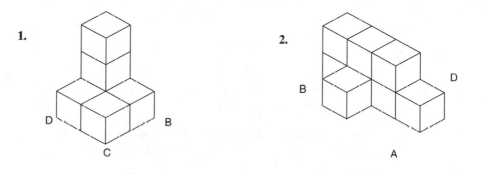

In Exercises 3 through 8, sketch the corner view of the building on isometric grid paper. You may wish to construct the buildings out of cubes first. Lines should be drawn only for edges of the buildings.

3.

3	3
2	1

D

4.

2	2	2
1	1	1

C

5.

2	2	
1	1	1
	1	

C

6.

3	3	
1	1	2
	1	

D

7.

	2	3
1	1	1
	1	

D

8.

	2
2	1
3	1

C

In Exercises 9 and 10, construct or visualize the building shown in the coded plan. Draw each corner view of the objects on isometric grid paper.

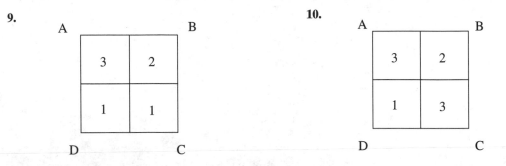

In Exercises 11 through 14, a coded plan and an indicated corner are given for each building. Sketch the building from that corner without the use of isometric grid paper.

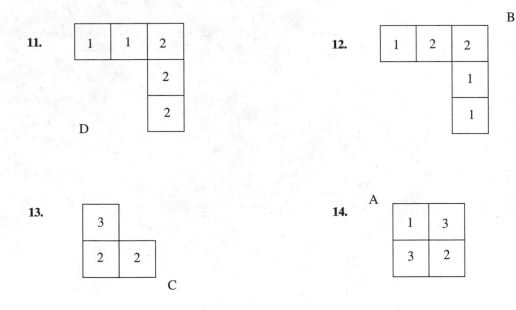

15. On isometric grid paper, draw two different views of a unit cube (i.e., a cube with $1 \times 1 \times 1$ dimensions). In the first view of the cube, the top should be visible. In the second view of the cube, the bottom should be visible.

16. A **wedge** is defined as a $2 \times 1 \times 1$ block cut in half diagonally. On isometric grid paper, draw at least two different views of a wedge.

3.3   OBLIQUE SKETCHING

Oblique sketching is another popular method for making pictorial sketches. With oblique sketching, one face of the object is parallel to the plane of the paper and the third dimension is shown as receding lines from this surface. The angle of the receding line is usually approximately 45°; however, this angle can be varied to show object features more clearly. Figure 3-9 shows an oblique sketch of the object shown as an isometric pictorial in

Figure 3-6, with the face defined by corners D and C shown parallel to the plane of the paper.

Coded Plan   Building   Oblique Sketch

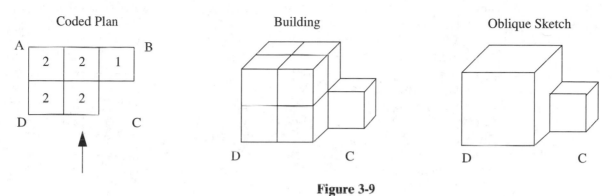

Figure 3-9

Notice that the two surfaces of this object which are parallel to the plane of the paper are shown in true shape and true size. The receding surfaces are seen distorted in an oblique pictorial; i.e., they appear as parallelograms. Figure 3-10 shows a cube drawn as both an isometric and an oblique pictorial.

Isometric Pictorial   Oblique Pictorial

Figure 3-10

As can be seen in this figure, in the isometric view of the cube, each face of the cube appears as a parallelogram, not as a square. In contrast, one face of the cube in the oblique pictorial appears as a square while the remaining surfaces appear as parallelograms. Thus, oblique sketching is sometimes preferred for objects that have one complicated surface because the complex surface can be shown in true shape, with the remaining surfaces shown distorted. This is especially true when drawing curved or irregular surfaces. With isometric sketches, the curved surface will always appear as an ellipse. If an oblique sketch is made, it is possible to create the drawing so that the curved surface appears as a circle. Figure 3-11 illustrates this phenomena. Sketching circular shapes in

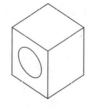
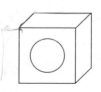

Isometric Pictorial   Oblique Pictorial

Figure 3-11

isometric views will be covered in Chapter 5. However, an oblique sketch of an object is not a true projection of the object whereas an isometric sketch is a true projection. For this reason, most computer software will not be able to display a three-dimensional object as an oblique pictorial.

When making oblique sketches, the length of the receding lines are not usually important. In fact, oblique sketches will typically look better if the receding lines are <u>not</u> shown in true length but are drawn shorter than true length. There are three types of oblique pictorials which are commonly made. A *cavalier* pictorial is one where the receding lines are shown true length; a *cabinet* pictorial is one where the receding lines are shown one-half of true length; and, a *general* pictorial is one where the receding lines are shown at any proportion other than true length or half-size. Figure 3-12 shows a cube drawn by each of these methods. Note that the appearance of the cube does indeed seem somewhat distorted in the cavalier pictorial where the receding length is shown true size.

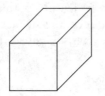

Cavalier Pictorial
Receding Length True Size

Cabinet Pictorial
Receding Length Half Size

General Pictorial
(Receding Length 1/4- Size)

Figure 3-12

When making oblique pictorials, you should choose the view of the object which has the most irregular features in it to be displayed in the plane of the paper. This will help to reduce the distortion of the object and will make the sketch easier to draw. Figure 3-13 shows an object in oblique pictorial which is drawn from two different vantage points. Notice that the first pictorial clearly shows what the object looks like with the least amount of distortion whereas, with the second pictorial, the amount of distortion for the primary features of the object may make it difficult to understand.

You can use square grid paper to help you create oblique sketches of objects. If you do this, however, recall from basic trigonometry that the length of a diagonal for a

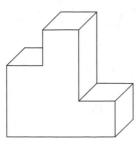 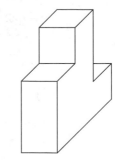

Pictorial with correct orientation   Pictorial with incorrect orientation

Figure 3-13

one-unit square is 1.4 units. Therefore, if you draw an oblique pictorial on square grid paper you will not be able to "count the squares" in the receding direction. If you do, your pictorial will appear even more distorted than it would if drawn as a cavalier pictorial. Figure 3-14 shows an object drawn on square grid paper. The first pictorial was drawn by counting grid squares along the diagonal in the receding direction, and the second pictorial was made by ignoring the grid spacing in the receding direction. The depth of the object is one unit in the direction perpendicular to the plane of the page. Notice that the second drawing looks much less distorted than the first.

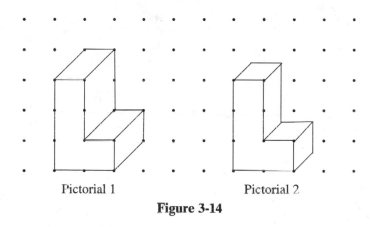

Pictorial 1   Pictorial 2

Figure 3-14

To draw an oblique pictorial from a coded plan, it is best to start with the first surface which is parallel to the plane of the paper. Draw this surface true shape and size, using grid paper if you need to. You then sketch the receding lines which extend back from this surface and draw the next surface (or part of a surface) which is parallel to the plane of the paper. Continue this procedure until your pictorial sketch is complete. Figure 3-15 demonstrates this procedure.

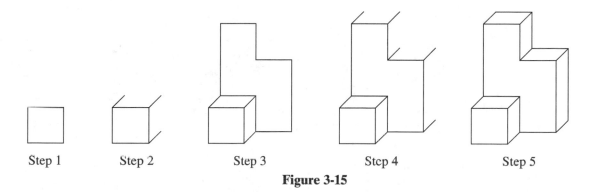

Step 1   Step 2   Step 3   Step 4   Step 5

Figure 3-15

EXERCISES 3.3

In Exercises 1 through 4, sketch a general oblique pictorial view with the face defined by the corners C and D parallel to the plane of the paper for the objects shown in coded plan.

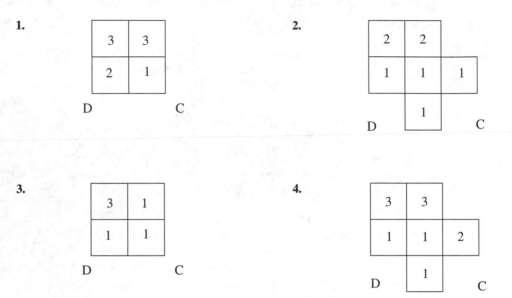

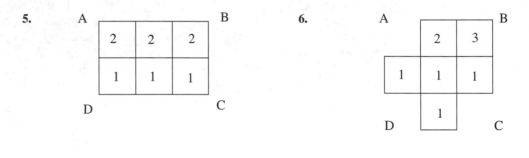

In Exercises 5 through 8, sketch an oblique pictorial view defined by corners A and D. Then sketch an oblique pictorial view defined by corners C and D. Which pictorial looks best?

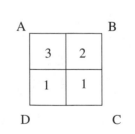

In Exercises 9 through 12, sketch both a cavalier and a cabinet pictorial for the objects defined in coded plan. Choose the orientation of the pictorial view so that the main object features are true size and shape.

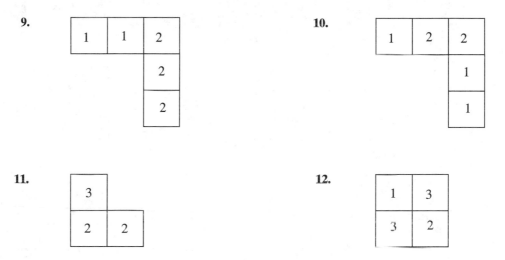

9.

1	1	2
		2
		2

10.

1	2	2
		1
		1

11.

3	
2	2

12.

1	3
3	2

3.4 LETTERING

Very few engineering drawings consist of only lines and curves. Most drawings contain notes, dimensions, labels, and other forms of annotation which are made up of letters and numbers. In previous times, an engineer's or a drafter's ability to correctly add the annotation to a drawing was critical to the clarity of the drawing for others who were interpreting the drawing. Eventually, lettering machines were developed so that accurate, clear, and uniform letters and numbers were easily achieved. These lettering machines were cumbersome to use, and annotation was often the most tedious part of creating engineering drawings. Fortunately, with the advent of computer-aided drafting and design software, drawing annotation using a lettering machine is more or less a "thing of the past."

As an engineer, you may be asked to provide a drafter, a contractor, or someone else with a sketch of your design ideas. In making these sketches, you will probably have to include either notes or dimensions on them. In order to clearly convey your design ideas, it is important that the annotation you include on the sketch is neat and legible. Sloppy lettering is often interpreted by others as sloppy work, which could lead to questions of credibility. By convention, drawing annotation uses all capitalized Gothic characters. Figure 3-16 shows the correct lettering technique to be followed when adding annotation to a sketch or a drawing. Note that most of the characters are made by moving the pencil horizontally from left to right and vertically from up to down. This is in agreement with good sketching practice as described earlier in this chapter. Recall that it is easier to sketch straight lines if you move the pencil from left to right and from up to down and if you keep your eye focused on the endpoint to which you are sketching.

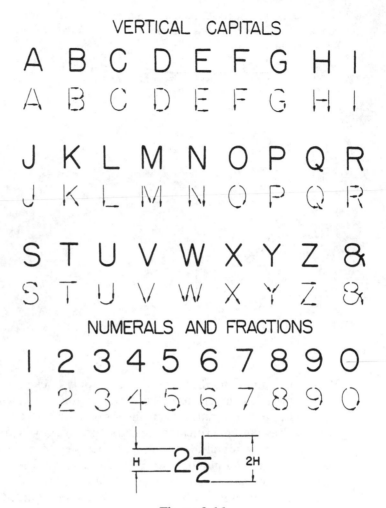

Figure 3-16

Your lettering will be much neater on your drawing if you first draw faint guidelines both on top of the lettering and beneath it on the paper. This will ensure that the characters and numbers are aligned on the paper and do not "drift" upwards or downwards. Figure 3-17 illustrates the use of a lettering guide line.

Lettering with a guide line

Lettering without a guide line

Figure 3-17

The lettering techniques illustrated in this section were developed for right-handed people. If you are left-handed, you will probably need to adjust some of the techniques in order to obtain neat and precise characters. Some left-handers will find it easier to make the lettering strokes from right to left just as in sketching.

EXERCISES 3.4

1. Practice writing all of the characters in the alphabet and all of the numbers (0-9) on an unlined sheet of paper. First sketch a faint guide line on the paper to aid in the alignment of the characters.

2. Write out the following title block using guide lines to align the text:

> YOUR NAME
> COURSE NUMBER
> SECTION NUMBER
> TODAY'S DATE

3.5 APPLICATION

In this section, you will learn about making schematic sketches. Although they are not 3-dimensional sketches, they are included in this chapter because they represent a type of sketch that you may be asked to make at some point in your engineering education or career. Schematic drawings are different from most of the engineering drawings you will encounter. They are comprised mainly of symbols that represent real, physical entities. Electrical circuit diagrams are an example of a schematic drawing. Figure 3-18 shows a simple circuit diagram with the type of symbols used in the drawing labeled. If you were an electrical engineer designing a circuit, you might start with a sketch of a circuit such as this during the preliminary stages of the design process.

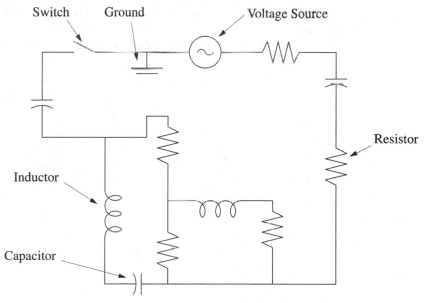

Figure 3-18

Another type of schematic drawing that you may find is a pipe instrumentation diagram as shown in Figure 3-19. These sketches are typically the first step in the design process for chemical engineers. The piping instrumentation diagram shows the location of valves and dial indicators in addition to the piping itself.

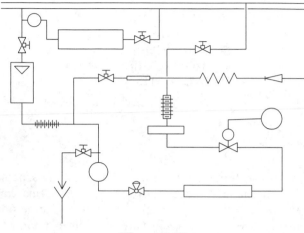

Figure 3-19

A final example of a schematic diagram is a computer programmer's flow chart, as shown in Figure 3-20. This diagram outlines the flow of information through a program and is typically the first step in algorithm development for computer programmers.

Schematic drawings typically show the "flow" of something through an engineered system. A circuit diagram shows the flow of current through an electrical circuit; a pipe diagram shows the flow of material through a process plant. Sometimes the "flow" is indicated on the schematic sketch with arrows or other symbols. There are no specific standards which govern the creation of schematic drawings, but there are some general rules of thumb you should follow when making this type of sketch. First, you should be familiar

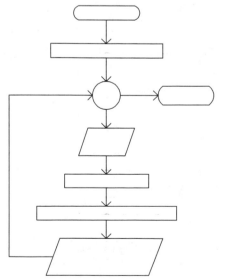

Figure 3-20

with the symbols used in the specific discipline. In other words, you should be able to convey your design ideas to others through the use of standard symbols which are understandable to anyone else who is familiar with the specific engineering discipline. Second, there is no specific size for each of the symbols, but they should be drawn so that they are proportionally correct with respect to one another.

EXERCISES 3.5

For Exercises 1 through 5, make hand-drawn sketches of the schematic drawings shown.

1.

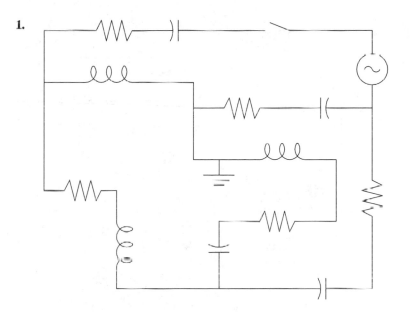

2.

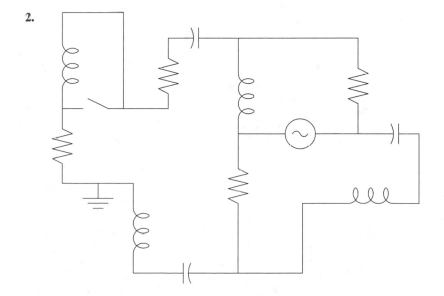

3.

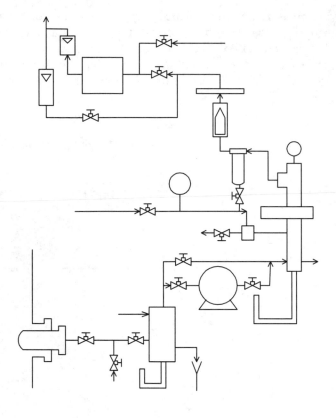

4.

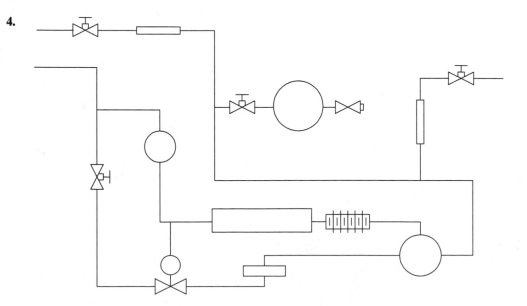

5.

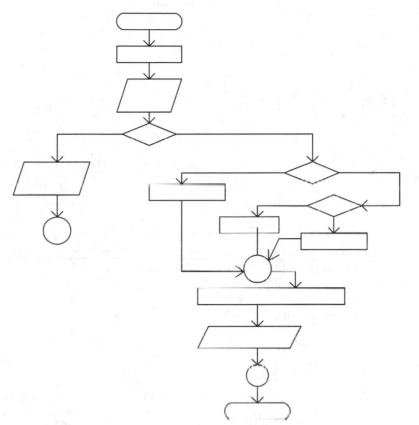

Transformation of 3-D Objects

• •

A geometric transformation of a solid is a one-to-one mapping of the points of a first solid object into the points of a second solid object. The second solid is called the **image** of the first under a given transformation. There are four types of geometric transformations that we will consider: a *translation*, a *dilation* (or scaling), a *rotation*, and a *reflection*. The rotation is the most difficult of the four geometric translations to visualize, so the greatest amount of time will be spent discussing it.

4.1 TRANSLATION OF OBJECTS

A geometric **translation** of an object is the sliding of the object in the X-, Y-, or Z-direction or in a combination of these directions. Figure 4-1 illustrates the translation of an object in each of these directions. Essentially, a translation takes place when an object is

Figure 4-1

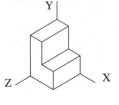

Original Object Position

Object Translated in
X-Direction Only

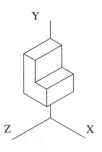

Object Translated in
Y-Direction Only

Object Translated in
X-Direction Only

"picked up" from one location and is moved to a new location without any turning of the object or physical changes to its shape. The image is merely the original object in a new location. In a translation, the isometric view of the image is identical to the corresponding isometric view of the original object. The edges of the object that started out parallel to the X-axis remain parallel to the X-axis after the object is moved, edges originally parallel to the Y-axis remain parallel to the Y-axis after the move, and edges originally parallel to the Z-axis remain parallel to the Z-axis after translation. Figure 4-2 shows an object before and after translation in the X, Y, and Z directions—notice that the isometric views of the object are identical.

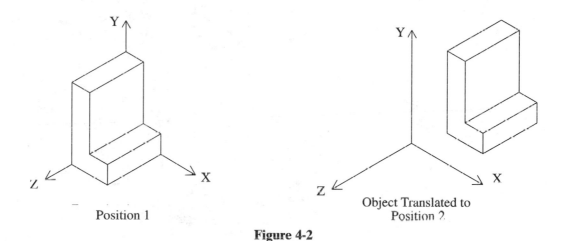

Position 1

Object Translated to
Position 2

Figure 4-2

Isometric grid paper enables you to be more quantitative about the translation of objects. Consider the L-shaped object shown before and after translation in Figure 4-3. When it is translated 2 units in the X-direction, −3 units in the Y-direction, and 1 unit in the Z-direction, the image appears as shown in the figure. One way of thinking about this is to think of moving *critical points*, e.g., vertices of the object, and then use the images of these points to sketch a translated image of the object. This is possible because each point undergoes the same translation that the solid as a whole experiences (e.g., A is translated to A', B is translated to B', and so on).

Note also that when you are making translations of an object in 3-D space, it is important to determine the exact X, Y, Z coordinates of the critical points. Recall from section 2-3 that appearances can be deceiving when portraying 3-D space on a 2-D piece of paper.

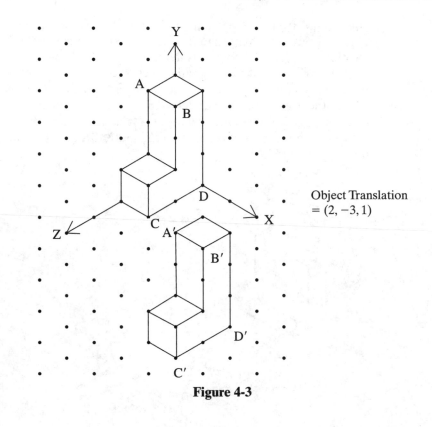

Figure 4-3

Object Translation
= (2, −3, 1)

EXERCISES 4.1

In Exercises 1 through 6, translate the object shown below as indicated in each exercise. Draw the image of each object after translation on isometric dot paper. On each drawing, indicate the location of the X-, Y-, and Z-coordinate axes.

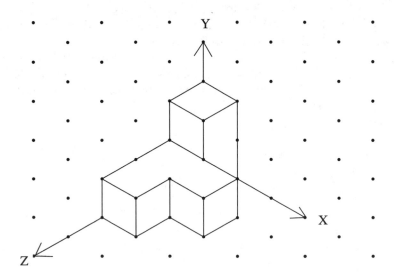

1. Translate the object 2 units in the X-direction.
2. Translate the object −3 units in the Y-direction.
3. Translate the object 1 unit in the Y-direction and 2 units in the Z-direction.
4. Translate the object −3 units in the X-direction and 2 units in the Y-direction.
5. Translate the object 4 units in the X-direction, −1 unit in the Y-direction, and 2 units in the Z-direction.
6. Translate the object −1 unit in the X-direction, 3 units in the Y-direction, and −2 units in the Z-direction.
7. What translation would be required to move the object shown below from its current location to a new location where point A would lie on the X-axis, point B would lie on the Y-axis, and point C would lie on the Z-axis? Assume point A currently has coordinates (5, 0, −2).

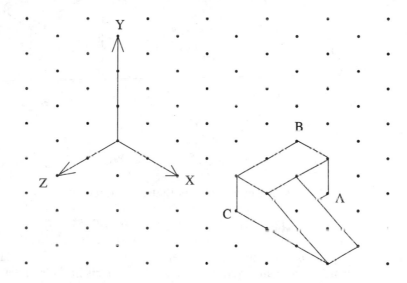

8. What translation would be required to move the object shown below from its current location to a new location where point A would lie on the X-axis, point B would lie on the Y-axis, and point C would lie on the Z-axis? Assume point B currently has coordinates (−5, 1, 1).

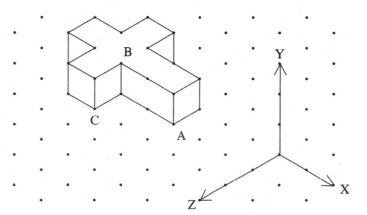

4.2 DILATION OF OBJECTS

A geometric **dilation** of an object with a multiplier $r \neq 1$ is an enlargement of the object if $r > 1$ and a shrinking of the object if $r < 1$. Consider the unit cube on the left in Figure 4-4. A dilation of the cube with $r = 3$ appears at the right. We say that the cube is enlarged or *dilated* by a **factor** (or multiplier) of 3. Isometric views of the object and image are similar but not identical as shown in this figure.

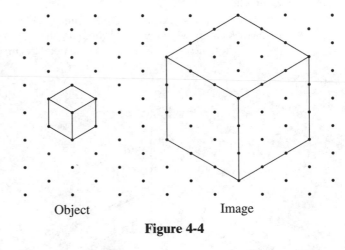

Object   Image

Figure 4-4

There are many objects that are similar in shape but different in size. Dry cereal boxes are one example. A Corn Flakes™ box with $12'' \times 9'' \times 3''$ dimensions would be similar in shape to a Grape Nuts™ box with $8'' \times 6'' \times 2''$ dimensions. In this case, the Grape Nuts™ box could be regarded as a Corn Flakes™ box reduced by a factor of $\frac{2}{3}$. Pieces of clothing or pairs of shoes that come in different sizes but the same style illustrate a practical application for scaling objects in the retail world.

EXERCISES 4.2

1. A triangular prism is shown below. Reduce the prism to $\frac{1}{3}$ its current size and make an isometric sketch of the reduced image.

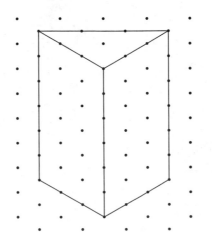

In Exercises 2 through 5, sketch an isometric view of the object shown in coded plan both before and after scaling by a factor of 2.

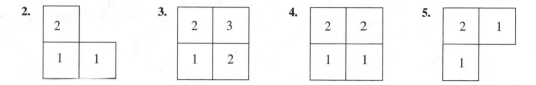

4.3 ROTATION OF OBJECTS ABOUT A SINGLE AXIS

A geometric **rotation** of an object is a turning of the object about a straight line that is referred to as the **axis of rotation**. In this section, we will discuss rotations occurring about a single axis, such as the X-, Y-, or Z-axis.

Imagine a door as it is moved from a position of closed to open, and let the vertical line along which the hinges lie be called the Y-axis (see Figure 4-5). As the door is pushed inward from a position of closed to a position of open, it is said to rotate about the Y-axis. Note that in rotating a 3-D object, the object remains in contact with the axis of rotation. If initially you are looking at the front of the door, then after a 90° counterclockwise rotation, you will be looking at an edge of the door. Figure 4-6 shows two views of a door before and after rotation about the Y-axis.

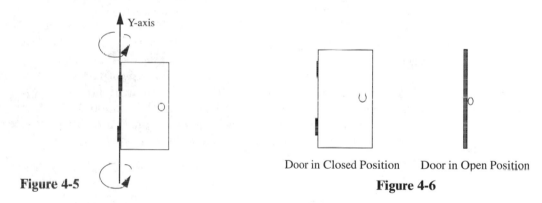

Door in Closed Position   Door in Open Position

Figure 4-5   **Figure 4-6**

The direction of rotation of a solid object is determined by the right hand rule. If the thumb of your right hand points along the axis of rotation, then your fingers will curl in the direction that the object rotates. In the case of the door which rotates about its hinges (shown in Figures 4-5 and 4-6), the thumb of your right hand points in the positive Y-direction, and the fingers of your right hand curl in the direction the door rotates (see Figure 4-7).

Figure 4-7

Figure 4-8 illustrates the rotation of a solid about each of the X-, Y-, and Z-axes. In each case, it should be noted that the solid object remains in contact with the axis of rotation. One of the object edges remains stationary while the rest of the object pivots about the axis. This is in contrast with the case of translation of solid objects where none of the object edges remain in their original position.

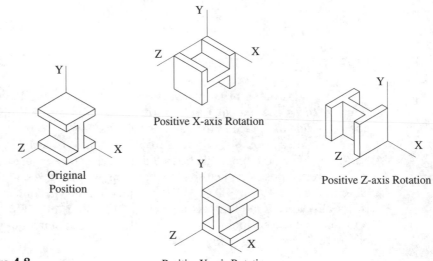

Figure 4-8

Original Position

Positive X-axis Rotation

Positive Y-axis Rotation

Positive Z-axis Rotation

An object can rotate in two directions about an axis. If the thumb of the right hand is pointed along the positive direction of an axis, then the fingers of the right hand curl in one direction. We will call such object rotation a **positive (or counterclockwise) rotation**. In contrast, if the thumb of the right hand is pointed along the negative direction of an axis, then the fingers of the right hand curl in the opposite direction. We will call such object rotation a **negative (or clockwise) rotation**. Figure 4-9 illustrates a positive as well as a negative rotation of an object about the X-axis.

Figure 4-9

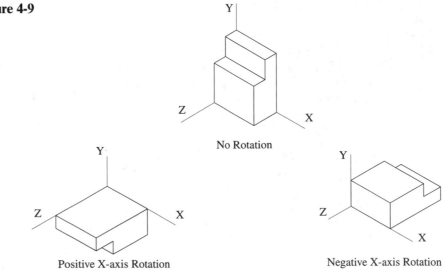

No Rotation

Positive X-axis Rotation

Negative X-axis Rotation

Another difference between the rotation and the translation of solids is that rotation will significantly change the appearance of the object in its isometric view. With rotation of solids about an axis, the edges originally parallel to the axis of rotation remain parallel after movement. However, the edges of the object parallel to axes other than the axis of rotation are no longer parallel to those same axes. This results in an isometric view of the object unlike that of the object in its original position. Figure 4-10 illustrates the rotation of an object about the Y-axis and the corresponding isometric views for the two orientations.

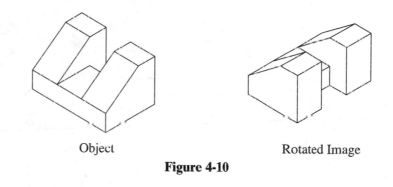

Object   Rotated Image

Figure 4-10

The following arrow coding scheme can be used to record the actual rotation(s) an object has experienced. Let a single arrow to the right (———→) represent a 90° positive rotation, a double arrow to the right (====→) represent a 180° positive rotation, a single arrow to the left (←———) represent a 90° negative rotation, and so on. The axis of rotation is then indicated to the right of the arrow(s). Therefore, the notation (———→ Y) would represent a 90° positive rotation about the Y-axis.

Imagine a unit cube constructed by folding up the flat pattern shown in Figure 4-11. This cube will spell the word CUBE as you proceed around each face. The letter H is on top of the cube, and the letter T is located on its bottom. We will choose an original position for the cube that exposes the faces lettered C, U, and H and then perform various rotations about the Y-axis. A sampling of possible rotations and the arrow coding of these

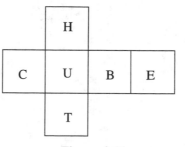

Figure 4-11

rotations are shown in Figure 4-12. The rotation in part c is a 270° negative (clockwise) rotation about the Y-axis and is equivalent to the 90° positive (counterclockwise) rotation shown in part a.

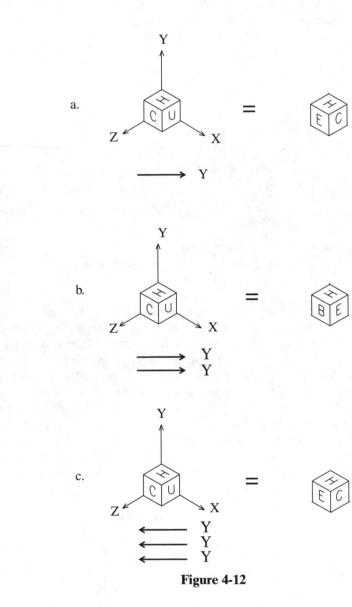

Figure 4-12

EXERCISES 4.3

In Exercises 1 through 4, rotate the given object as indicated, draw the isometric view of the object in its final position on isometric grid paper, and indicate the arrow coding of the rotation below your drawing.

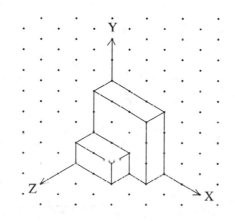

1. A negative 90° rotation about the X-axis.
2. A positive 90° rotation about the Z-axis.
3. A positive 90° rotation about the Y-axis.
4. A positive 270° rotation about the Y-axis.

In Exercises 5 and 6, describe in writing and with arrow coding what rotation the object has experienced about the indicated axis. Express each answer using as few arrows as possible.

5.

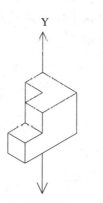
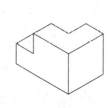

Initial Position   Final Position

6.

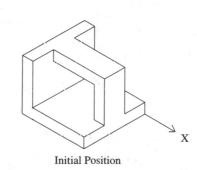
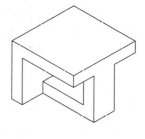

Initial Position   Final Position

In Exercises 7 through 10, indicate whether the object rotations are equivalent.

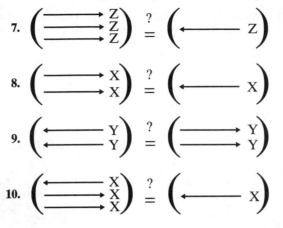

7. $\left(\begin{array}{l} \longrightarrow Z \\ \longrightarrow Z \\ \longrightarrow Z \end{array}\right) \stackrel{?}{=} \left(\longleftarrow Z\right)$

8. $\left(\begin{array}{l} \longrightarrow X \\ \longrightarrow X \end{array}\right) \stackrel{?}{=} \left(\longleftarrow X\right)$

9. $\left(\begin{array}{l} \longleftarrow Y \\ \longleftarrow Y \end{array}\right) \stackrel{?}{=} \left(\begin{array}{l} \longrightarrow Y \\ \longrightarrow Y \end{array}\right)$

10. $\left(\begin{array}{l} \longrightarrow X \\ \longrightarrow X \\ \longrightarrow X \end{array}\right) \stackrel{?}{=} \left(\longleftarrow X\right)$

4.4   ROTATION OF OBJECTS ABOUT TWO OR MORE AXES

In the same manner that objects are rotated about a single axis, they can be rotated about two or more axes in a series of steps. In Figure 4-13, an L-shaped object has a 90° negative rotation about the X-axis followed by a 90° positive rotation about the Z-axis. The arrow coding for such a 2-step rotation would be $\left(\begin{array}{l} \longleftarrow X \\ \longrightarrow Z \end{array}\right)$. Figure 4-14 shows an object that has had a 90° negative rotation about the X-axis followed by a 90° positive rotation about the Y-axis. The arrow coding for this 2-step rotation would be $\left(\begin{array}{l} \longleftarrow X \\ \longrightarrow Y \end{array}\right)$. When a rotation occurs about a single axis, an entire edge remains in its original position. When rotations occur about two different axes, only a single pivot point remains stationary. In Figure 4-14, this single pivot point is the origin (x = 0, y = 0, z = 0).

When a solid rotates about two different axes, the final location and orientation of the solid depend on the order in which the rotations are performed. That is, rotation of a

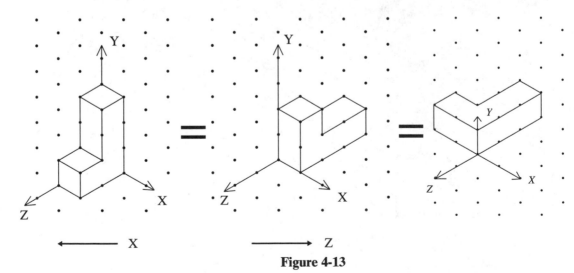

Figure 4-13

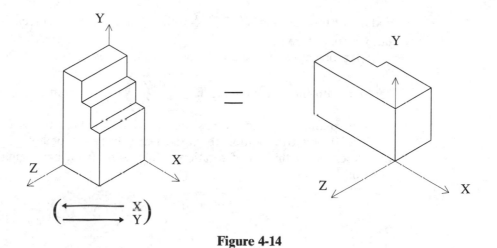

Figure 4-14

solid about two different axes is not commutative. In other words, rotation about X followed by rotation about Y is not the same as rotation about Y followed by rotation about X. Figure 4-15 illustrates the failure of the commutative property in object rotation about two different axes. In the first case, the original object was rotated about the X-axis and then about the Y-axis. The arrow coding for this rotation is $\left(\begin{array}{c} \longrightarrow \\ \longrightarrow \end{array} {\small \begin{array}{c} X \\ Y \end{array}} \right)$. In the second case, the original object was rotated first about the Y-axis and then about the X-axis. The arrow

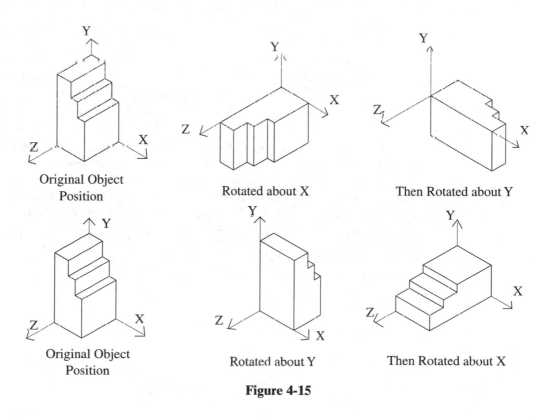

Original Object Position   Rotated about X   Then Rotated about Y

Original Object Position   Rotated about Y   Then Rotated about X

Figure 4-15

coding for this rotation is $\left(\Longrightarrow \begin{smallmatrix} Y \\ X \end{smallmatrix} \right)$. Note that in these two cases the objects' final orientations are different.

An object can be rotated about as many axes as you wish. For example, an object rotation with the arrow coding $\left(\begin{smallmatrix} \longrightarrow & X \\ \Longrightarrow & Z \\ \longrightarrow & Z \\ \longleftarrow & Y \end{smallmatrix} \right)$ means that the original object has a 90° positive rotation about the X-axis; then a 180° positive rotation about the Z-axis; and, last of all, a 90° negative rotation about the Y-axis. The L-shaped object shown in Figure 4-16 has experienced this set of rotations while going from its original position to its final position.

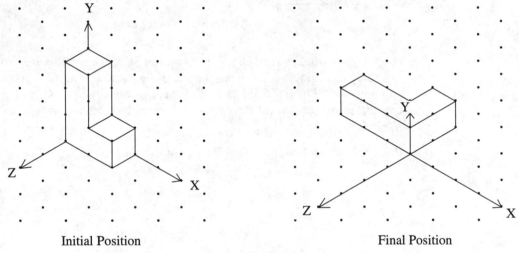

Initial Position   Final Position

Figure 4-16

A set of object rotations about two or three axes can sometimes be reduced to a simpler set of object rotations (i.e., the arrow coding has fewer arrows in it, but the end result is the same). It is not always easy to find a simpler set of rotations. You need to concentrate on the initial and final positions of the object and find the shortest set of rotations between these two positions. A simpler set of object rotations for the rotation shown in Figure 4-16 would be $\left(\rightleftarrows \begin{smallmatrix} Z \\ X \end{smallmatrix} \right)$, a 90° positive rotation about the Z-axis followed by a 90° negative rotation about the X-axis.

EXERCISES 4.4

In Exercises 1 through 4, rotate the solid below as indicated. Then, on isometric dot paper, draw the object in its final position relative to the X-, Y-, and Z-axes, and indicate the arrow coding of the rotation below your drawing.

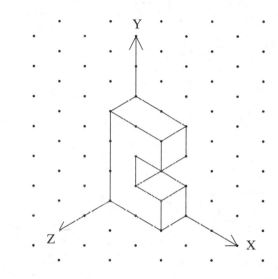

1. A positive 90° rotation about the X-axis followed by a positive 90° rotation about the Z-axis.
2. A negative 90° rotation about the X-axis followed by a positive 180° rotation about the Y-axis.
3. A positive 180° rotation about the Z-axis followed by a positive 90° rotation about the Y-axis.
4. A negative 90° rotation about the Z-axis followed by a positive 180° rotation about the X-axis.

In Exercises 5 through 8, study how the object on the top line is rotated [Hint: It may be helpful to identify the axis of rotation or to devise a coding scheme for the rotation]. Then rotate the object below it in the same manner and draw the second object in its final position on isometric dot paper.

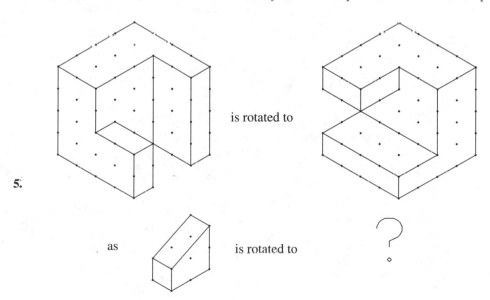

5.

as   is rotated to   ?

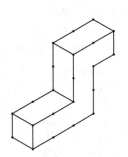   is rotated to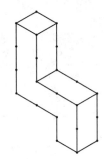

6.

as   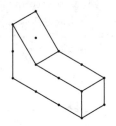   is rotated to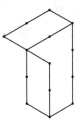

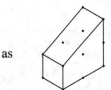   is rotated to

7.

as   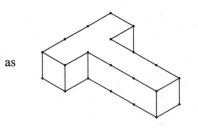   is rotated to

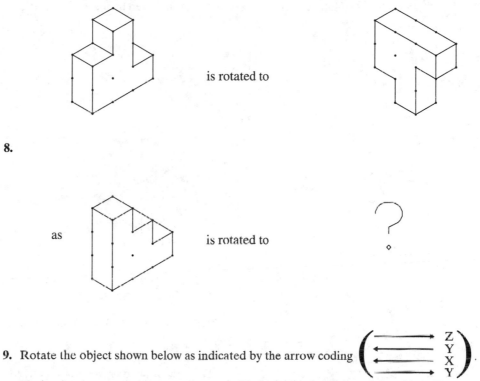

8.

as 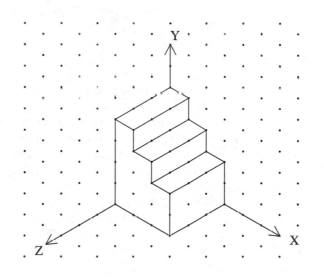 **is rotated to** ?

9. Rotate the object shown below as indicated by the arrow coding $\left(\begin{array}{c} \longrightarrow \; Z \\ \longleftarrow \; Y \\ \longleftarrow \; X \\ \longrightarrow \; Y \end{array} \right)$.

Find a simpler set of object rotations to achieve the final position of the object. Express your answer using arrow coding. Are other answers possible?

10. Rotate the object shown below as indicated by the arrow coding.

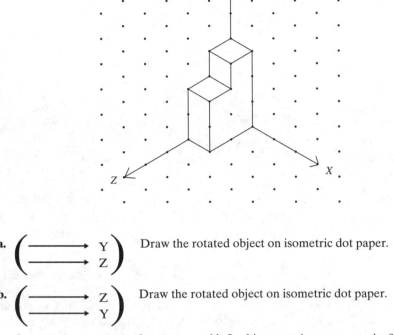

a. $\left(\begin{array}{c} \longrightarrow\ Y \\ \longrightarrow\ Z \end{array} \right)$ Draw the rotated object on isometric dot paper.

b. $\left(\begin{array}{c} \longrightarrow\ Z \\ \longrightarrow\ Y \end{array} \right)$ Draw the rotated object on isometric dot paper.

c. Compare your answers for parts a and b. Is object rotation commutative?

4.5 REFLECTION OF OBJECTS

A **reflection** of an object across a plane P is a geometric transformation that associates each point A of the object with an image point A′ such that plane P is a perpendicular bisector of the line segment AA′. (Note: If a point of the object lies on plane P, then it is its own reflection.) Figure 4-17 shows two different objects and their reflections across plane P. Plane P is technically not a mirror, but a plane through which the object is reflected. Hence, the image of the reflection and the original object are located on opposite sides of plane P.

A 3-D object is said to be **symmetrical** if a plane can cut the object such that the part of the object on one side of the plane is the reflection of the part of the object on the other side of the plane (or vice versa). In Figure 4-17, the T-shaped object is symmetrical, but the other object is not. Figure 4-18 shows the T-shaped object being cut by plane Q such that the left side of the object is a reflection of the right side of the object. It is important to note that a plane of symmetry exists in one single object whereas a reflection creates two separate objects—the original object and its reflected image.

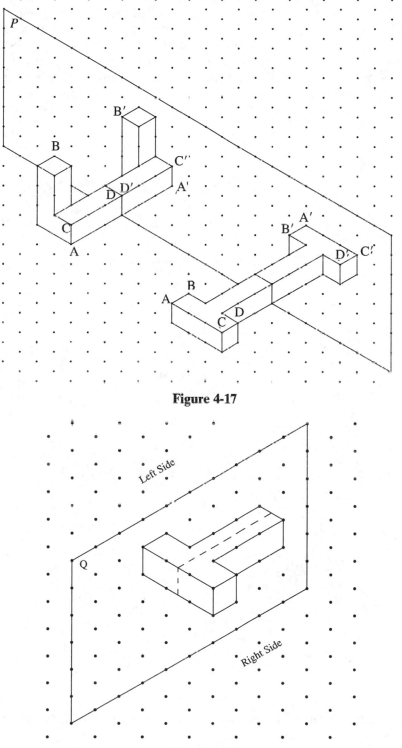

Figure 4-17

Figure 4-18

For symmetrical 3-D objects, it is possible to use two rotations to achieve the same effect as a single reflection, provided the plane of symmetry and the plane of reflection are perpendicular to one another. In this case, the axis of rotation must be the line formed by the intersection of these two planes. Figure 4-19 shows the reflection of a symmetrical object, the plane of symmetry Q, the plane of reflection P, and the resulting axis of rotation. Figure 4-20 shows the object in Figure 4-19 experiencing two positive 90° rotations about the axis of rotation. Notice that the images of the object after the geometrical transformations of reflection (Figure 4-19) and rotation (Figure 4-20) are the same.

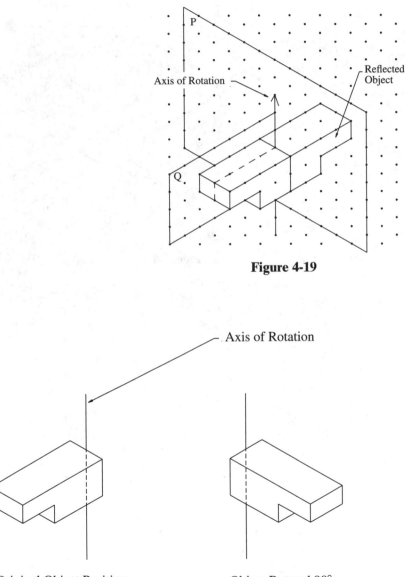

Figure 4-19

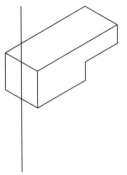

Original Object Position   Object Rotated 90°   Object Rotated 180°
                                                                          (Reflected Object)

Figure 4-20

EXERCISES 4.5

In Exercises 1 through 4, copy plane P on a piece of isometric dot paper and draw the reflection of the given object. (Note that in Exercise 3, the object is located 2 units away from plane P.)

1.

2.

3.

4.

Note: This object is located 2 units away from the plane P.

5. How many planes of symmetry does the object pictured in Exercise 3 have?
6. How many planes of symmetry does the object pictured in Exercise 4 have?
7. How many planes of symmetry does the object pictured in Exercise 2 have?

8. On isometric grid paper, copy plane P and the object shown in the figure below. Draw the reflection of the object through plane P. Can the reflection be accomplished by an equivalent set of two 90° rotations? If so, draw the axis of rotation in its proper location.

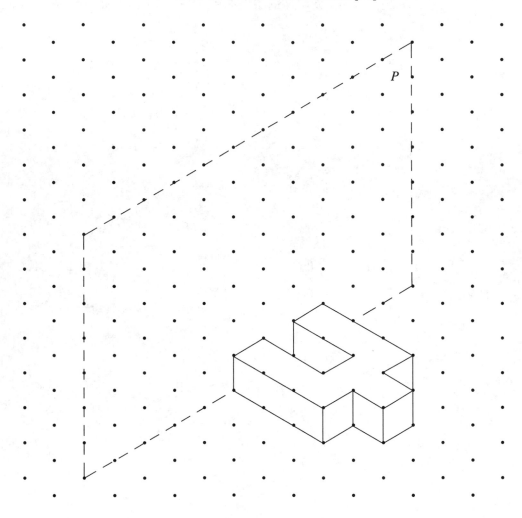

4.6 APPLICATION

Objects are typically made up of surfaces which intersect to form edges. Sometimes it is useful to create a flat pattern for a 3-D object so that all of the surfaces on the object lie in the same plane. Thus, each of the surfaces is rotated about an edge until it lies in the same plane as the adjoining surface. Surfaces can be rotated one at a time until all of the surfaces defining the object lie in the plane. When a flat pattern is created for an object, the edges (where surfaces intersect) can also be referred to as fold lines. The fold lines are the axis of rotation for the surfaces defining the object. Imagine a cube with the word CUBE spelled out on four of its sides, with one letter per side. Figure 4-21 gives an isometric view of such a cube along with two possible flat patterns for such a cube. In these flat patterns, the fold lines are represented by solid lines. Imagine rotating the surfaces about

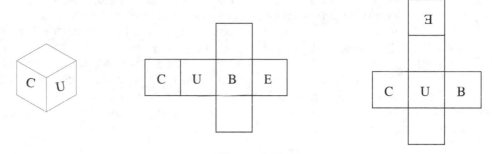

Figure 4-21

their fold lines until the cube is formed. Creating the flat pattern which can be folded to obtain a 3-D object is also called *developing* a pattern. Thus, the flat pattern is often referred to as a pattern development.

One application of creating flat pattern developments is in the area of sheet metal working. Sheet metal is typically made out of thin (less than $\frac{1}{4}$") galvanized steel. One of its primary uses in the industrial setting is for duct-work, which carries heat and air conditioning through buildings. Straight, regular-shaped duct-work is available in standard sizes. However, in creating all the duct-work necessary in a building, custom-made pieces are sometimes necessary for going around corners, changing directions, or providing a transition from one size of duct-work to another. In order to make these specialized 3-D pieces out of a flat sheet of metal, pattern developments are created. These developments are made true size so they can be placed directly on the sheet metal and cut out. Tabs are usually placed at the seam line of the pattern development so the final product can be put together easily. In the case of sheet metal working, tabs are usually slipped into joints and screwed to other surfaces to add strength to the metal piece. Besides providing the general shape for cutting the sheet metal, the pattern is also used to scribe lines on the metal showing where it is to be folded. The final 3-D product is achieved by folding the cut sheet metal into the desired shape. Once the piece is folded, the seam line is fastened using the tabs. Because duct-work is usually open at each end to allow for the free flow of air through a building, typical sheet metal pattern developments are open-ended.

Figure 4-22 shows an isometric view of a 3-D shape that requires a flat pattern development. For this application, however, you will typically work with a top and front

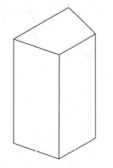

Figure 4-22

view of the object as shown in Figure 4-23. The top and fornt views of the object are known as orthographic views. Orthographic views will be described in detail in Chapter 5. In order to create the flat pattern development for this open-ended object, you should first imagine drawing a line all the way around the perimeter of the object. This line is usually around the bottom of the object and is called the **stretch-out line**. You then draw this stretch-out line to the right of the front view as one continuous line segment as illustrated in Figure 4-23. The true size of the perimeter can be measured in the top view. The lengths of the segments as viewed from the top are true length because they are parallel to the top view.

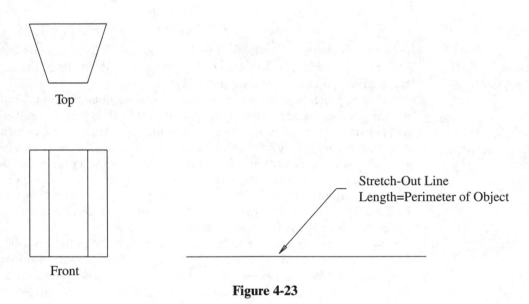

Figure 4-23

A vertical edge on the object is chosen as the seam line for the development. Typically, the seam line is chosen as the shortest edge. In this case, all edges are the same length, therefore the choice of a seam line is arbitrary. The addition of the tab sections will not be shown in this example. The fold lines for this development are the corners of the object, and the flat pattern development will also include the locations of the fold lines. Note that when viewed from above, the object corners appear as points, and when viewed from the front, the corners appear as straight vertical lines. Each object corner is now labeled in both the front and the top view. In labeling the object corners, both endpoints of the corner are given. In the top view, the two point labels coincide. This is illustrated in Figure 4-24.

The bottom points (A, B, C, and D) are positioned appropriately along the stretch-out line so that the lengths of the line segments (i.e., AB, BC, CD, and DA) are shown true size. The true lengths of the line segments are obtained in the top view of the object. Figure 4-25 shows the placement of these marks on the stretch-out line.

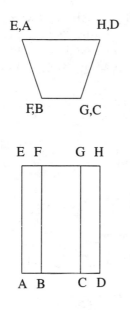

Figure 4-24

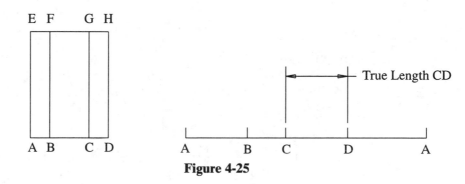

True Length CD

Figure 4-25

Note that the development along the stretch-out line begins and ends with the same point. To complete the pattern development, the fold lines (corners) are drawn perpendicular to the stretch-out line. The length of a fold line is equal to the height of the object at each corner. This true size height is seen in the front view. Figure 4-26 shows the completed flat pattern development for the object.

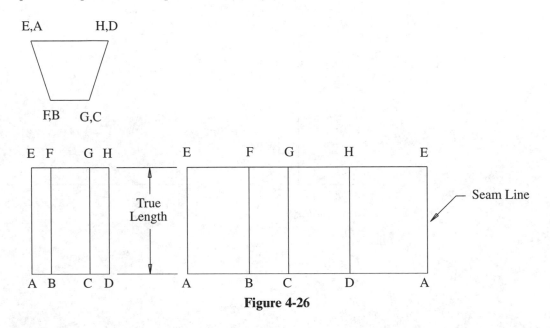

Figure 4-26

Figure 4-27 shows a flat pattern development for a folded sheet metal part that is not of a uniform height. The procedure used in creating this pattern development is identical to the one outlined in the previous example. Note that in this figure, the development starts and ends at point D. This ensures that the shortest fold line is the seam line. Figure 4-28 is an isometric view of the object created by folding the pattern shown in Figure 4-27.

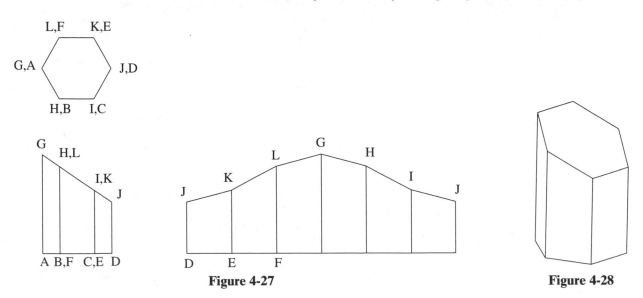

Figure 4-27

Figure 4-28

EXERCISES 4.6

Using square grid paper, sketch the flat pattern developments for the sheet metal parts shown in
Exercises 1 and 2. Top and front orthographic views are given for each sheet metal part. Estimate
the lengths of the sides along the inclined surfaces.

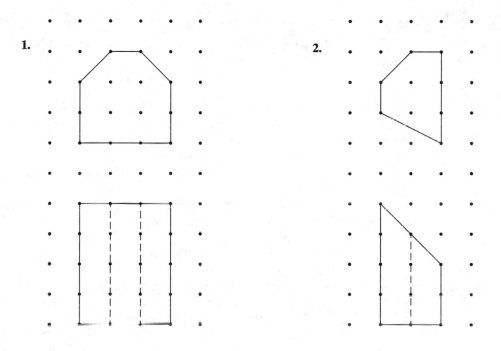

1.

2.

C H A P T E R 5
Orthographic Projection

. . . .

Isometric views are useful for showing a 3-D representation of a solid object on a flat (2-D) piece of paper. However, there are many instances where isometric sketches do not adequately display information in an understandable form. As an example of this, consider a house. An isometric view of the house would show qualitatively what the house looked like from the outside. If, however, you wanted to show the interior of the house and the layout of the rooms on each floor, an isometric view would be particularly confusing because the walls from the rooms would run into each other and overlap. Thus, in order to display the layout of the rooms, an orthographic or plan view is normally used. A plan view is drawn as if you are located at a point in space directly above a floor of the house and are looking down on it. In this way, you are able to accurately view the sizes and relative locations of all the rooms on a particular floor of the building.

This illustrates an important feature in creating views of a building or an object. In general, the object remains fixed in space, and you "move" around the object in order to "see" what it looks like from different angular perspectives. For example, an isometric view is created as if you are looking straight down a line which is the extension of a diagonal connecting opposite corners of a cube. By looking down the diagonal, you see all three dimensions (height, width, and depth) of the object. In a plan view of a house, the observer is located at a point directly above the floor—the house remains stationary. From this direction, only two dimensions, the width and the depth of the house, are visible—the height is not.

In general, views where only two dimensions are shown are called orthographic views. Typically, the top view (also referred to as plan view) is one where the object's width and depth are displayed; the front view (also known as the front elevation) shows the width and height of an object, and the side view (or side elevation) shows the depth and the height. Figure 5-1 illustrates the relationship between dimensions and each ortho-

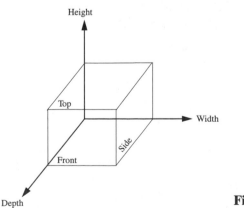

Figure 5-1

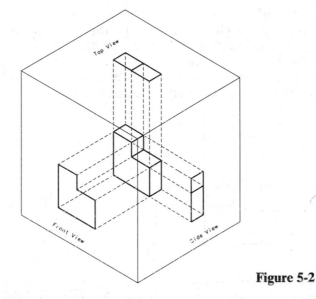

Figure 5-2

graphic view. Note that any two views will display all the necessary dimensional information for an object, but frequently a third view will be included for purposes of clarity.

Figure 5-2 shows how orthographic views are derived from the geometry of a 3-D object. In creating orthographic views, you can imagine the object is enclosed within a glass cube. Each of the orthographic views is created by projecting edges of the object onto the panes of the glass cube. The edges are projected such that the lines of projection (or projectors) are perpendicular to the panes of the glass (the term orthographic comes from the Greek *ortho*, which means perpendicular) and parallel to each other. The projectors create a profile of the front and side views on vertical panes of the cube and a profile of the top on a horizontal pane. To obtain the standard layout of the views on a flat surface such as a sheet of paper, it is common practice to "unfold" the box as shown is Figure 5-3.

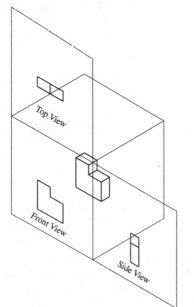

Figure 5-3

In the system of orthographic projection, there are six principal views corresponding to the six panes of the glass cube. These principal views are the front, top, back, bottom, left side, and right side. Figure 5-4 shows the standard arrangement of the top, front, and right side views (with projectors shown as dashed lines) as they would appear in an engineering drawing. Note that in this system of orthographic projection, the edges from one view project perpendicularly to the adjacent views.

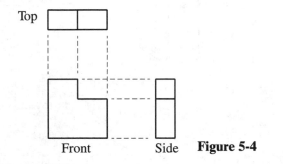

Front   Side   **Figure 5-4**

As in writing, where rules of grammar are needed for effective written communication, rules for the construction of orthographic drawings are needed for effective graphical communication. These rules are often referred to as standards. One of the rules or standards to be followed is that orthographic views of an object should be aligned with one another. Figure 5-5 illustrates properly and improperly aligned orthographic views. You will learn of other rules or standards as we further examine orthographic projection.

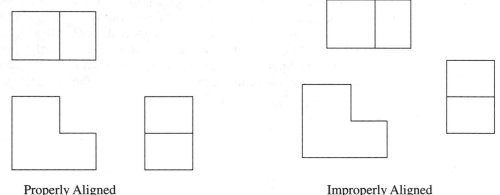

Properly Aligned   Improperly Aligned
Orthographic Views   Orthographic Views

Figure 5-5

5.1 NORMAL SURFACES

Engineering drawings are created as representations of real world, three-dimensional objects. Real objects consist of surfaces which intersect in edges. Depending upon line-of-sight, in orthographic drawings those surfaces and edges will appear as bounded areas, as lines or as points. Throughout this text we will use the terms "surface" and "edge" when referring to real world features and "area," "point," and "line" when referring to how those features are represented in a drawing.

In the case of objects built out of blocks, all surfaces on the object are normal surfaces. Normal surfaces are defined as those which are parallel to one of the six glass panes of the cube and hence perpendicular or normal to the projectors to that pane. You should note that a surface parallel to the top pane is perpendicular to the front, back, and side panes; a surface parallel to the front pane is perpendicular to the side, top, and bottom panes and so on. A normal surface parallel to the front pane would appear as an area in the front view, a vertical line in the side views, and a horizontal line in the top view. Figure 5-6 illustrates this idea.

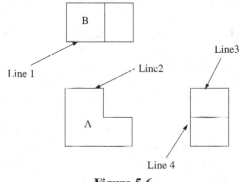

Figure 5-6

In this figure, surface A is parallel to the front view and is seen there in its true size and shape and appears as line 4 in the right side view and line 1 in the top view. Similarly surface B is parallel to the top view and appears as a true size area in the top view and as line 2 in the front and as line 3 in the right side.

One advantage of using orthographic views to describe an object is that the normal surfaces will appear true size and shape, and included angles true size, in at least one of the views. Thus, you are able to measure dimensional lengths and the size of angles directly. This is not the case in isometric drawings. As an example of this, consider the case of a cube. In reality we know that all faces of the cube meet at right angles. However, when an isometric view of a cube is constructed, if you were to measure the angle between the faces, that angle would be either 60° or 120° depending upon the pair of faces chosen. In a set of orthographic views of the same cube, angles between adjacent sides measure 90°. The reason for this is that in the orthographic drawing, the faces of the cube, being normal surfaces, are parallel and perpendicular to the viewing planes. In the isometric drawing, none of the faces are parallel to the viewing plane. Figure 5-7 graphically illustrates this concept.

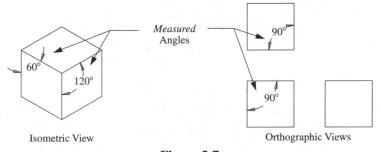

Figure 5-7

When creating an orthographic drawing or sketch, the common practice is to use lines to represent the edges of the object. Edges occur when two surfaces intersect. Sometimes when constructing an engineering drawing, there are edges of an object which would be hidden from a particular viewing angle and hence in particular views. We do not wish to omit these edges as they can provide valuable graphical information regarding the object. Instead, another standard of engineering graphics is applied. The hidden edges are drawn as dashed lines rather than the continuous line used for the representation of visible edges. In practice, the dashed lines are referred to as hidden lines and the continuous lines are referred to as object lines. According to the standards of engineering graphics, if an object line coincides with a hidden line, only the object line is shown.

Two orthographic views of an object provide complete dimensional information about that object since each view displays two dimensions. For example, a top view will show the width and the depth and a front view will show the width and height, so all three dimensions are represented within these two views and the object is completely defined. In many cases, because of the complexity of individual features, an object will be accurately defined by two orthographic views if and only if those two views are carefully chosen. The two views should include the most details in visible edges and show the most features in profile. This is referred to as choosing the most descriptive views.

Figure 5-8 shows five views of a simple object along with an isometric view of the same object. Note the use of the hidden line standard in the left and bottom views. Can you choose two descriptive views by which to represent this object?

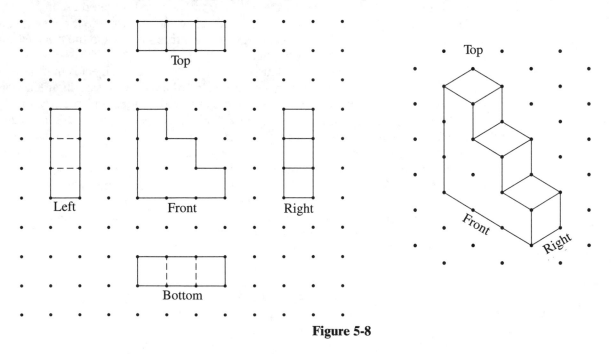

Figure 5-8

Hidden Line Intersections. Many times when you are creating orthographic views of objects you will have a situation where hidden lines intersect either other hidden lines or object (solid) lines. If this is the case, there are drawing conventions which govern hidden line intersections which you should follow in order to achieve clear and precise design representation. Figure 5-9 shows the conventions which apply when a hidden line intersects a solid object line or two object lines. In each of these cases, it is important to show the exact extent of the object and hidden lines. When a hidden line and an object line intersect in a "T," it is important to show the exact location of the start of the hidden line, and thus, the first dash of the hidden line should start at the object line. When the hidden line intersects in an "L," a gap is shown between the object lines and the first dash of the hidden line. If the dash started at the corner, then it would not be clear exactly how long the object line was. Ambiguity such as this should be avoided for clear graphical communication. When a hidden line crosses an object line, that hidden line should straddle the object line with a gap, not with a dash.

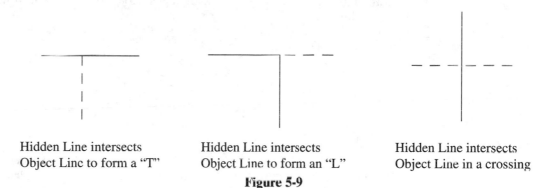

Hidden Line intersects
Object Line to form a "T"

Hidden Line intersects
Object Line to form an "L"

Hidden Line intersects
Object Line in a crossing

Figure 5-9

These hidden line conventions are illustrated in Figure 5-10 which shows the front, top, and right side views of an object. In this figure, the intersections labeled "A" show hidden lines intersecting an object line to form a T, intersections labeled "B" form an L, and the intersection labeled "C" shows a crossing of a hidden line with an object line.

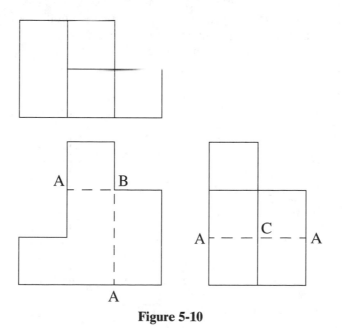

Figure 5-10

When hidden lines intersect other hidden lines, clarity concerning the exact extent of the lines is once again the driving factor. Figure 5-11 shows some of the drawing conventions which have been established for this type of hidden line intersection. When a hidden line intersects another hidden line in a "T," then the dash of one hidden line should start in the middle of a dash from the other hidden line. In this way, the exact location of the start of the hidden line is clearly defined. Similarly, if two hidden lines intersect to form an "L" then the point of interest is the exact location of the corner between the two lines. Therefore, the corner is joined between the two hidden lines. When two hidden lines intersect in a crossing, then the hidden line which defines the feature which is in front of the other is drawn with a dash at the intersection. In the hidden line crossing shown in Figure 5-11, the vertical hidden line is depicted as being in front of the horizontal hidden line.

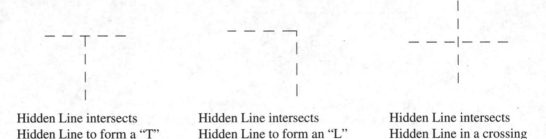

Hidden Line intersects   Hidden Line intersects   Hidden Line intersects
Hidden Line to form a "T"   Hidden Line to form an "L"   Hidden Line in a crossing

Figure 5-11

Figure 5-12 shows the front, top, and right side views of an object which illustrates these hidden line intersection conventions. In this figure, the intersection labeled "A" shows hidden lines intersecting to form a T, the intersection labeled "B" form an L, and the intersection labeled "C" shows a crossing of a two hidden lines. Note that, for the hidden line crossing, the vertical hidden line is in front of the horizontal hidden line and thus has a dash at the intersection instead of a gap. Note also the hidden line intersections with object lines in this figure.

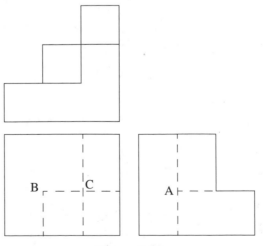

Figure 5-12

One common occurrence of hidden line intersections on engineering drawings is in illustrating drilled or tapped holes which do not extend all the way through the object (blind holes). In this case, because of the machining process, a conical shape is formed at the end of the cylindrical hole. Figure 5-13 shows how a blind hole is depicted on an engineering drawing. Notice how in each of the corners of this hidden feature, the dashes of the hidden lines form distinct points.

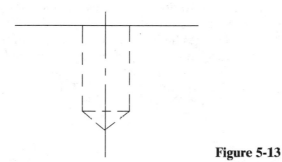

Figure 5-13

The hidden line intersections discussed in this section have developed over the years in an effort to achieve clarity on engineering drawings. If you are sketching, you will have no trouble in maintaining these conventions. If you are using computer-aided drafting software, however, you may have to adjust some of your drawing techniques or some line lengths in order to meet these standards. Unfortunately, most drafting software will create dashes and spaces according to an internal algorithm which automatically sets their lengths and may not correspond to conventional practice.

3-D View Construction. A common technique used for the enhancement of visualization skills is to present the student with two orthographic views of an object and have them construct a third view. Since the two given views provide a full set of dimensional information for the object, the solution involves an interpretation of the given views. This involves techniques which will be used by the engineer in reading and checking engineering drawings for information and accuracy. One such technique involves examining individual features of the object and visualizing them as they would appear in the solution view. For example, suppose the two views (top and front) of the object shown in Figure 5-14 were given and a third view (right side) was required. The top view

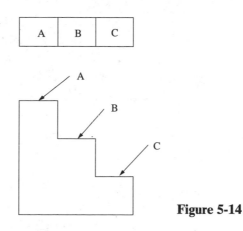

Figure 5-14

shows the width and the depth and the front view shows the height and the width. The side view to be constructed will show the height and the depth. The height dimension will be projected directly from the front view to the side view and the depth dimension will be transferred from the top.

The surfaces which appear as visible areas in the top view have been labeled A, B, and C. These surfaces will appear as horizontal lines in the front and side views because they are normal surfaces. To draw the representations of these surfaces in the side view, project them horizontally into that view from the front view. This projection gives the height at which the features lie. The length of the lines in the side view is obtained from the depth dimension shown in the top view. This is depicted in Figure 5-15. The single vis-

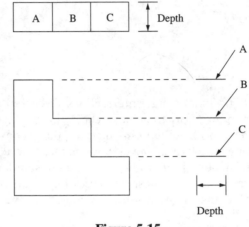

Figure 5-15

ible bounded area shown in the front view is the area view of a normal surface. This will appear as a horizontal line in the top view and a vertical line in the side view. In this example, you can see that the back surface is identical in size and shape to the front surface, otherwise, hidden lines would be visible in the front view. Figure 5-16 shows the complete drawing of the three views of this object.

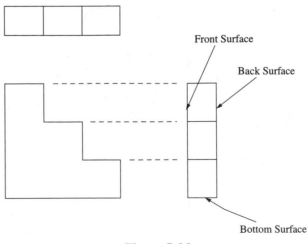

Figure 5-16

Constructing Isometrics from Orthographic Views. The construction of an isometric view may be performed using the three given orthographic views. This can be accomplished by using the box method. When creating an isometric by this method, first sketch the bounding box for the object on grid paper. The dimensions of the bounding box are the overall dimensions (height, width, and depth) of the object as seen in the orthographic views. This step is shown in Figure 5-17.

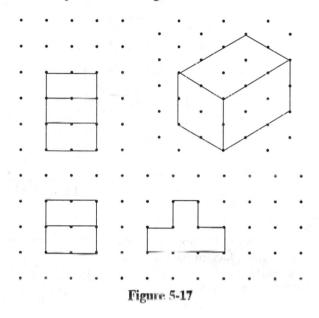

Figure 5-17

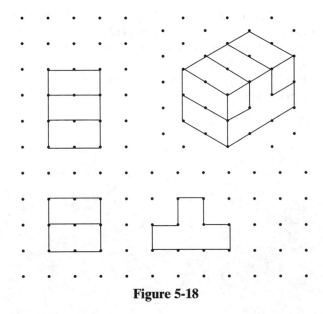

Figure 5-18

After sketching the bounding box, draw each of the orthographic views on the corresponding face of the box, as illustrated in Figure 5-18. Add and remove lines until all of the surfaces of the object are depicted. This process results in the final isometric view shown in Figure 5-19.

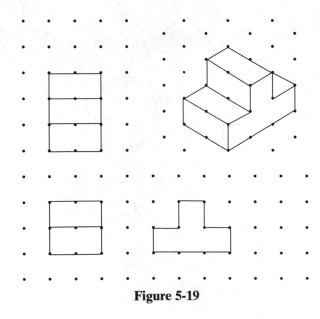

Figure 5-19

Surface Identification.   Another type of problem in orthographic projection, which aids in the visualization of objects, is surface identification. Recall that objects are made up of surfaces which enclose a volume of space. Each surface is visible in all views of the object—sometimes as an area and sometimes as a line. The vertices which define the surfaces project orthographically from one view to another. Figure 5-20 shows the top and front views of an object with all of the vertices which define the surfaces of the object labeled.

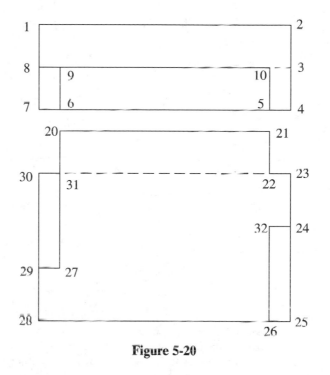

Figure 5-20

Figure 5-21 shows an isometric pictorial of the object defined in Figure 5-20. If you look in the front view, the surface which is defined by points 30 and 23 in the front view is defined by points 1, 2, 4, 5, 10, and 8 in the top view. For the surface defined by points 9 and 6 in the top view, which points define the surface in the front view? This is a normal surface which is perpendicular to both the front and the top views, so it is seen as a line in the front view and is defined by points 20 and 27. What about the edge 10, 5 seen in the top view? This edge is seen as a single point in the front view—point 21. Conversely, the edge defined at point 25 in the front view is seen as the edge 3, 2 in the top view.

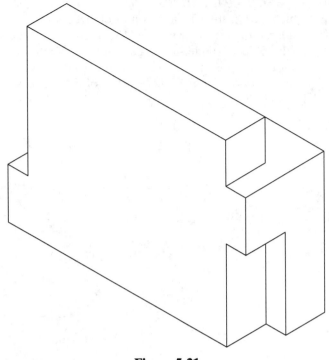

Figure 5-21

EXERCISES 5.1

1. Sketch the figure shown below using conventional practice to depict the hidden line intersections. Note: most of the intersections shown in this figure are <u>not</u> currently drawn by conventional practice. When hidden lines cross each other, assume that the vertical line is in front.

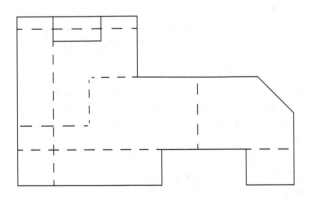

In Exercises 2–5 sketch the top, front, and right side views on square grid paper. Assume there are no hidden cubes.

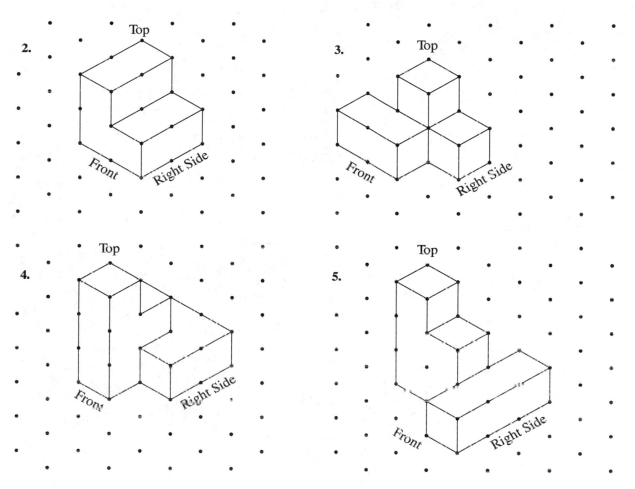

2.

3.

4.

5.

In Exercises 6–11, sketch the figures on square grid paper and add a minimum number of missing lines to complete the three views of the objects shown.

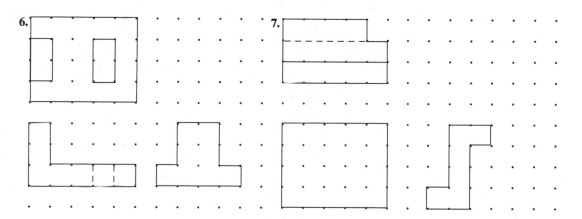

6.

7.

8.

9.

10.

11..

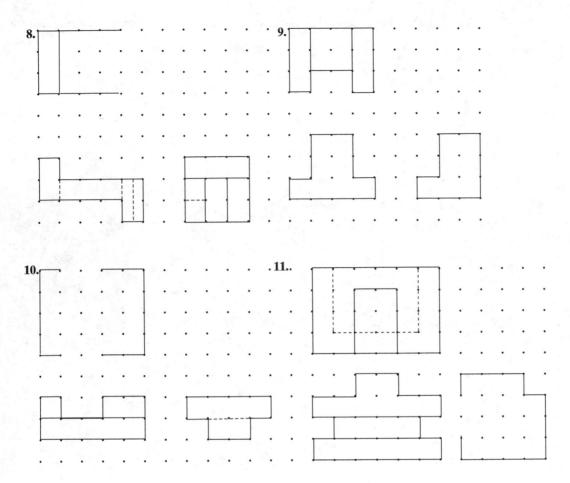

In Exercises 12 and 13, complete the table to identify the surfaces and edges in the opposite view. Make sure that you identify a surface as a surface and an edge as an edge.

12.

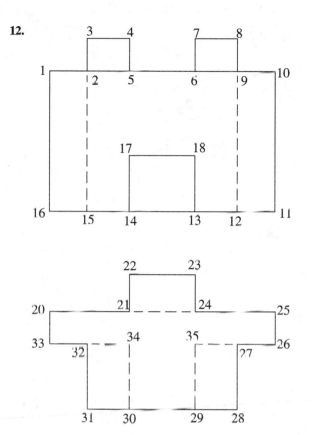

6, 7	Surface	
1, 10	Surface	
23, 24	Surface	
22, 23, 24, 21	Surface	
20, 25	Edge	
27	Edge	
31	Edge	
3, 4	Edge	

13.

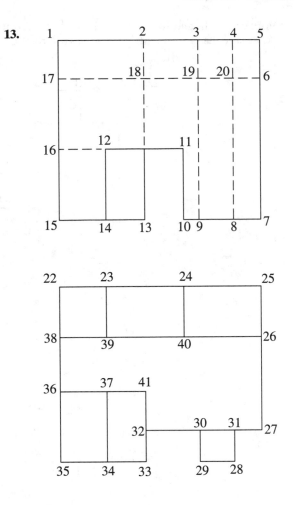

17, 6	Surface	
36, 37, 34, 35	Surface	
41, 33	Surface	
35, 33	Surface	
12	Edge	
30, 31	Edge	
5	Edge	
38, 26	Edge	

For Exercises 14–16, the complete top and front views of an object are given. Copy the two views onto square grid paper and then construct the right side view for each. On a separate sheet of isometric grid paper, construct the isometric view of the object.

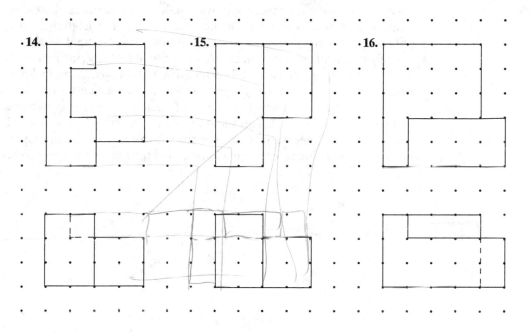

14.

15.

16.

5.2   INCLINED SURFACES

So far in our discussion of isometric and orthographic drawings, we have limited our consideration to normal surfaces only. Although normal surfaces alone can be used to demonstrate the principles of constructing orthographic and isometric views of an object, by observing your surroundings you can see that very few "real world" objects are made up entirely of normal surfaces. In contrast to normal surfaces, inclined surfaces are defined as follows: in a set of three mutually orthogonal viewing panes, an inclined surface is inclined with respect to two of the orthographic viewing planes and is perpendicular to the third. Thus, an inclined surface will appear as a foreshortened area in two of the views and as an inclined line in the third. You should note that, as opposed to normal surfaces, the area views of the inclined surface are not true size in any of the principal views. Figure 5-22 illustrates an inclined surface in an object and the corresponding projected views.

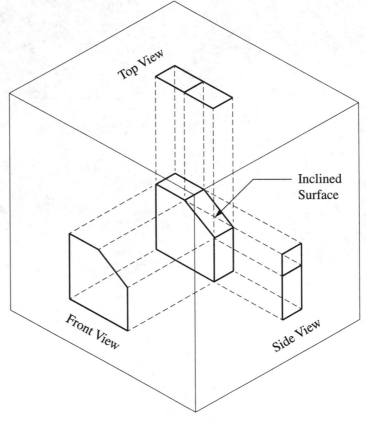

Figure 5-22

Figure 5-23 shows the orthographic views for the objects shown in Figures 5-2 and 5-22. As can be seen from this figure, the top and side views of both objects are identical, and the general shapes of the objects are best depicted in the corresponding front views. Note that these front views are absolutely necessary for the complete shape description of the objects: without them, you would not be able to tell exactly what the objects looked like. Further note that the size of the area view for the inclined surface is different in the right side view than in the top view. Neither of the views show the inclined surface as true size since in neither view is the surface parallel to the viewing plane (remember our definition of a normal surface). The size of the area views in both cases is smaller than the actual size of the surface. We say that the surface appears *foreshortened* in these two views.

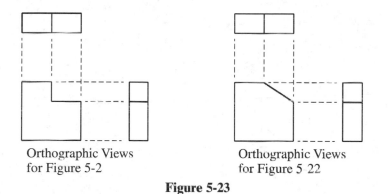

Orthographic Views
for Figure 5-2

Orthographic Views
for Figure 5 22

Figure 5-23

You should note that while the area views of an inclined surface are foreshortened, the basic shape is maintained. Right angle corners of the surface will appear as right angle corners in both of the foreshortened views. The number of edges which bound the area must be the same in all area views. Figure 5-24 depicts an object with an inclined surface labeled A. The corresponding top and side view of the surface are also indicated. Note that the two area views of the surface are the same general shape with the same number of sides and that right angle corners remain right angles but the relative size of the surface appears different in the two area views

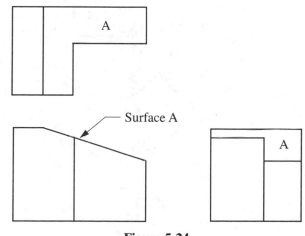

Surface A

Figure 5-24

Perhaps the easiest way to visualize this concept is to consider a door. With the door closed, your line of sight is normal to its surface and you can see the door true size and shape while standing directly in front of it. The door appears as a normal surface. When the door is opened part of the way, swinging on its hinges, you no longer see the door as true size. The surface of the door is inclined with respect to your line of sight: the door appears as an inclined surface. When the door is open all the way, you see the surface of the door as a line. This surface of the door now lies parallel to our line of sight and the door appears as the line view of a normal surface. This is illustrated in Figure 5-25.

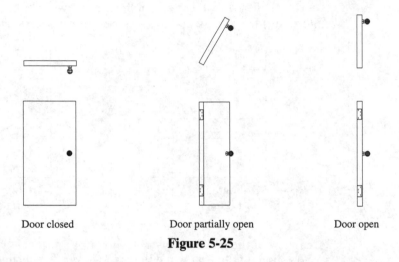

Door closed   Door partially open   Door open

Figure 5-25

When constructing an orthographic view of an inclined surface, it is frequently easiest to consider each vertex (edge to edge intersection) of the surface independently. You can then locate each vertex in each of the given views and project the vertices to the construction view. The vertices are connected to create edges in the same order as those in the other given area view. Figure 5-26 shows labels of the vertices for the inclined surface for the object shown in Figure 5-24.

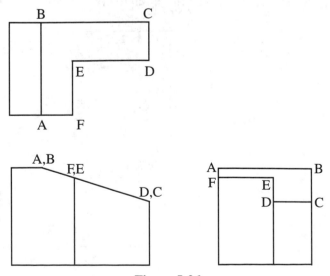

Figure 5-26

To construct an inclined surface in an isometric drawing, you typically locate the two endpoints of each inclined edge and draw a straight line between them. This will often be more easily accomplished after the normal surfaces of the object have been drawn. Figure 5-27 illustrates how inclined edges are located in isometric views and the final isometric drawing of the object.

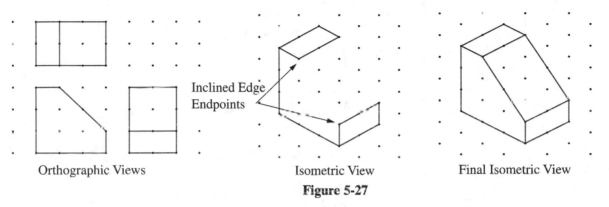

Orthographic Views   Isometric View   Final Isometric View

Figure 5-27

When constructing isometric views of objects that contain inclined surfaces, you should be careful to select an orientation of the object that makes the inclined surface appear as a visible area. Figure 5-28 shows two possible orientations of the object from Figure 5-27. As can be seen from Figure 5-28, the view labeled Isometric 1 clearly shows an area view of the inclined surface while the view labeled Isometric 2 does not. The second view does not present a clear representation of the appearance of the object. This ambiguity should be avoided to achieve effective graphical communication.

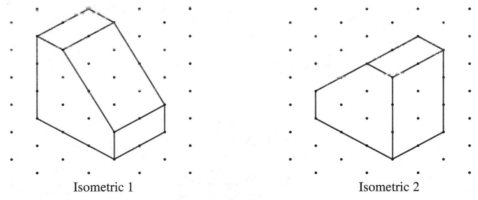

Isometric 1   Isometric 2

Figure 5-28

EXERCISES 5.2

In Exercises 1–4 sketch the top, front, and right side views of the objects on square grid paper.
Assume there are no hidden cubes.

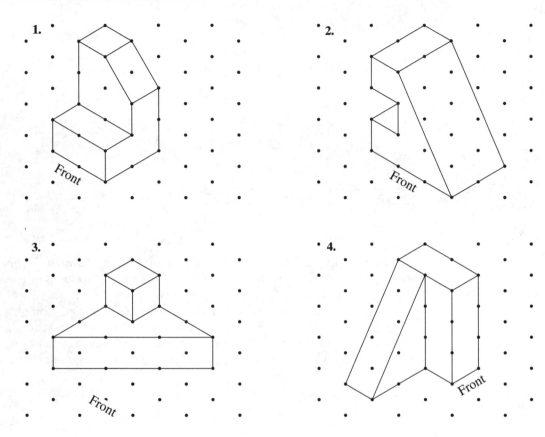

1.

Front

2.

Front

3.

Front

4.

Front

In Exercises 5–10, sketch the figures on square grid paper and add a minimum number of missing lines to complete the three views of the objects shown. Add lines to at most, two views.

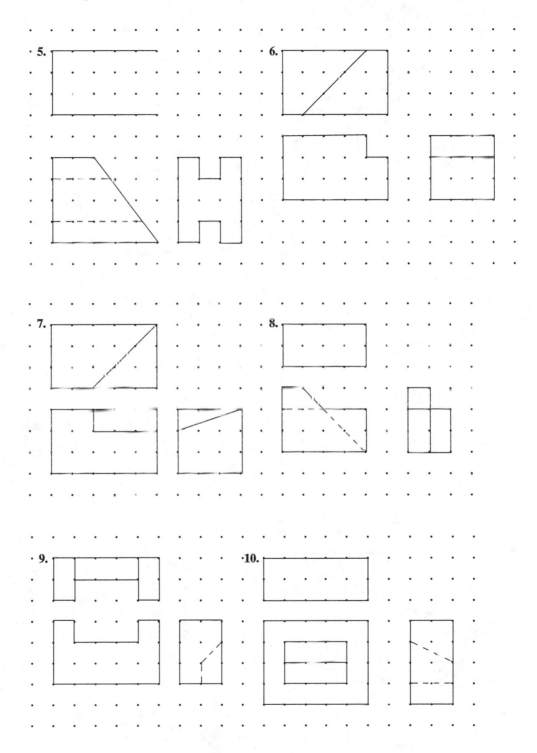

In Exercises 11 and 12, complete the table to identify the surfaces and edges in the opposite view. Make sure that you identify a surface as a surface and an edge as an edge.

11.

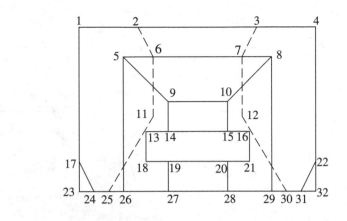

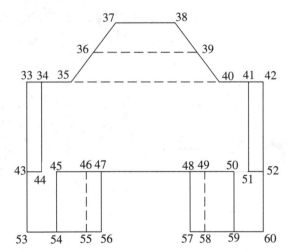

17, 24, 23	Surface	
33, 42	Surface	
35, 37	Surface	
12, 30	Surface	
51, 52	Edge	
35, 37	Edge	
2	Edge	
50	Edge	

12.

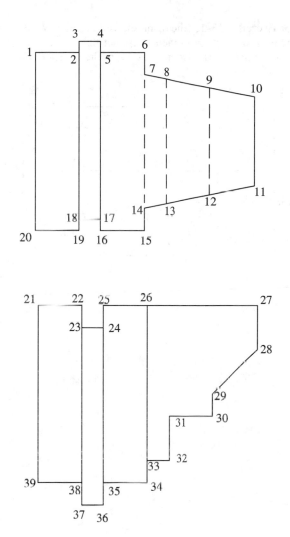

25, 26, 34, 35	Surface	
29, 30	Surface	
14, 11	Surface	
27, 28	Surface	
8, 9	Edge	
33	Edge	
18	Edge	
14, 15	Edge	

For Exercises 13–15, the front and right side views of an object are given. Copy the two views onto square grid paper and then construct the top view for each. On a separate sheet of isometric grid paper, construct the isometric view of the object.

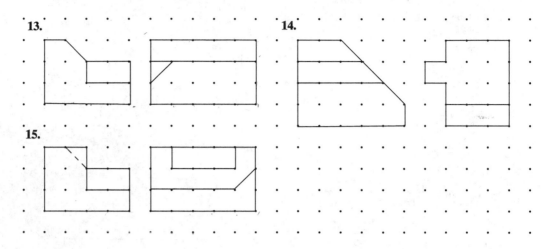

13.

14.

15.

16. In the problems shown below, the circles indicate the location of missing views. Select the correct view from the 30 proposed views shown. A view may be used more than one time.

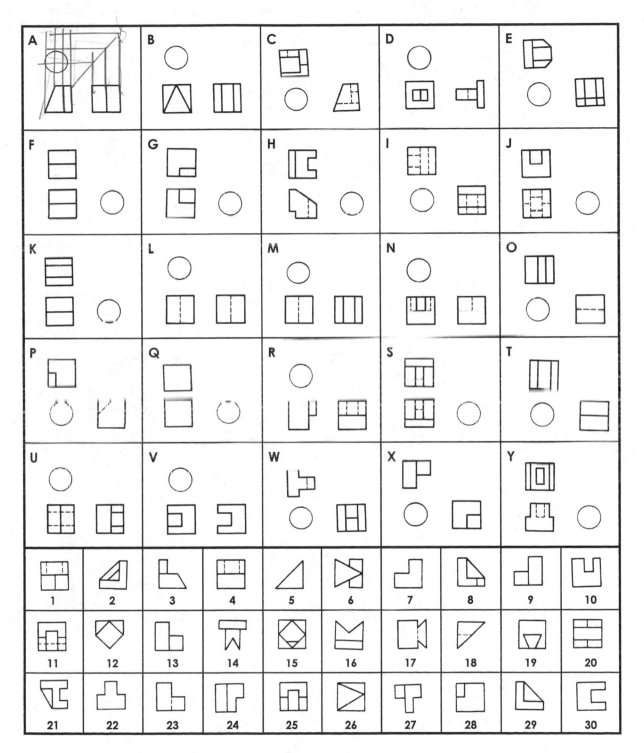

5.3 OBLIQUE SURFACES

An oblique surface is one whose orientation is neither parallel nor perpendicular to any of the six principal viewing planes. Therefore, the surface will appear as an area in all of the principal views (a surface must be perpendicular to a viewing plane to appear as a line). Since the surface is not parallel to any of the viewing planes, none of the area views will depict the surface as true size. Figure 5-29 shows the orthographic projection for a simple oblique surface. This surface was created by "cutting off" one corner of a block.

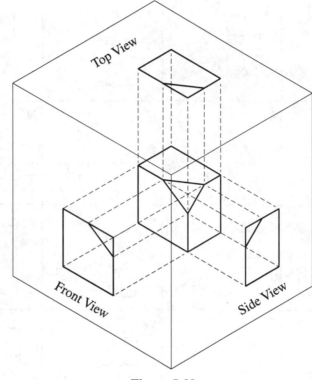

Figure 5-29

Oblique surfaces will be among the most difficult for you to visualize. Because of the general orientation of oblique surfaces, they may appear with a very different looking shape in different principal views. However, the number of edges and vertices for the surface must always remain the same. To create a third orthographic view of an oblique surface, it is often easiest to determine the location of the vertices defining the surface and "connect the dots." A similar procedure may be followed for the creation of an isometric view of an oblique surface. Figure 5-30 shows the procedure for creating the three orthographic views and an isometric view of an oblique surface. Note that the vertices in the views on the left have been numbered. You will find that this technique will make it easier to keep track of them. Since two views provide all three dimensions, this will guarantee that the edge exists.

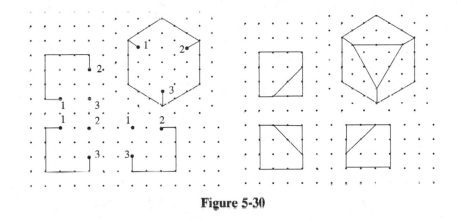

Figure 5-30

Figure 5-31 shows three orthographic views and an isometric view of a more complicated object containing an oblique surface. Look at the views of the object and verify the location of the vertices and the edges of the surface in each view. Notice also that the oblique surface is defined by the same number of edges and vertices in all views.

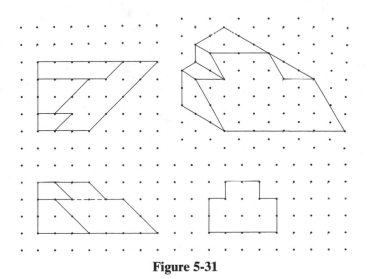

Figure 5-31

EXERCISES 5.3

In Exercises 1–3 sketch the top, front, and right side views of the objects on square grid paper. Assume there are no hidden cubes.

1.

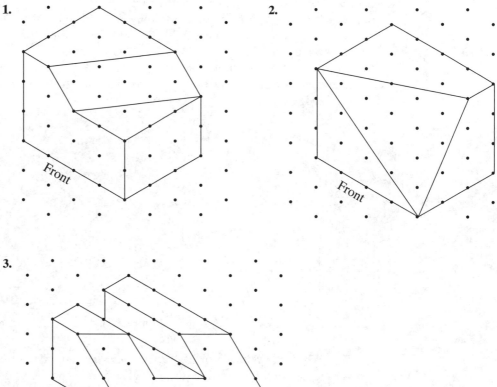

Front

2.

Front

3.

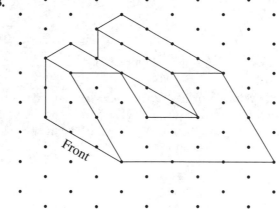

Front

In Exercises 4–9, sketch the figures on square grid paper and add a minimum number of missing lines to complete the three views of the objects shown. Add lines to at most, two views.

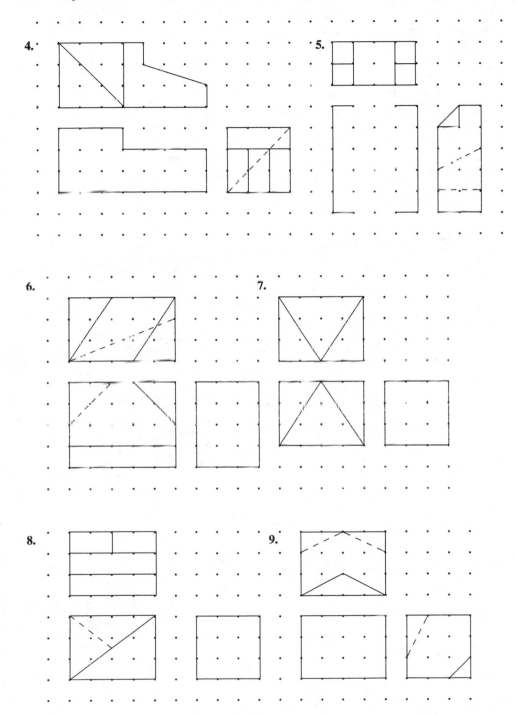

In Exercises 10 and 11, complete the table to identify the surfaces and edges in the opposite view. Make sure that you identify a surface as a surface and an edge as an edge.

10.

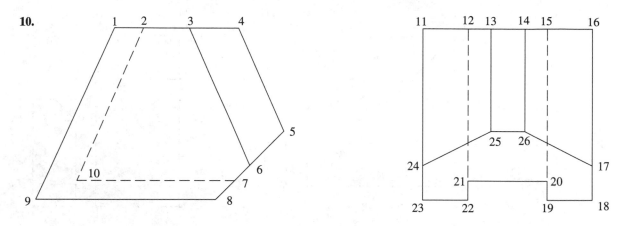

1, 9	Surface	
4, 5	Surface	
10, 7	Surface	
5, 8	Surface	
3, 6	Edge	
21, 22	Edge	
4	Edge	
7	Edge	

11.

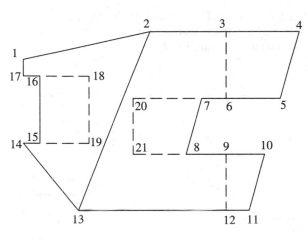

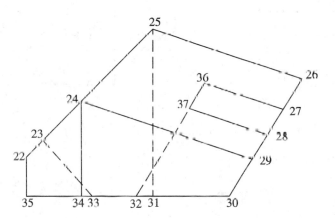

23, 33	Surface	
13, 14	Surface	
11, 13	Surface	
32, 36	Surface	
2, 13	Edge	
36, 27	Edge	
30	Edge	
14	Edge	

For Exercises 12–14, two views of an object are given. Copy the two views onto square grid paper and then construct the third view for each. On a separate sheet of isometric grid paper, construct the isometric view of the object.

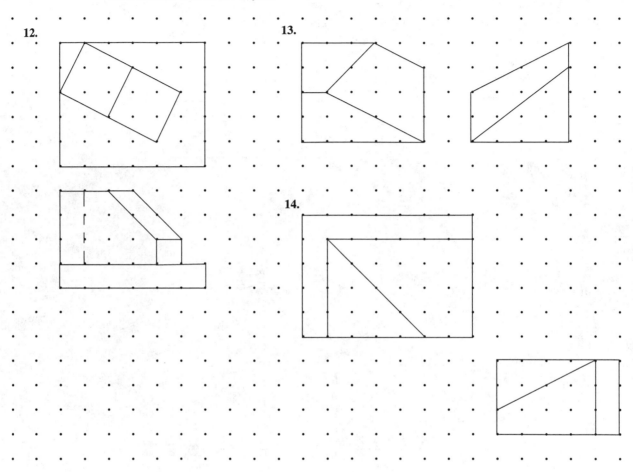

12.

13.

14.

For Exercises 15–24, select the pictorial view of the object which will produce the orthographic views that are given.

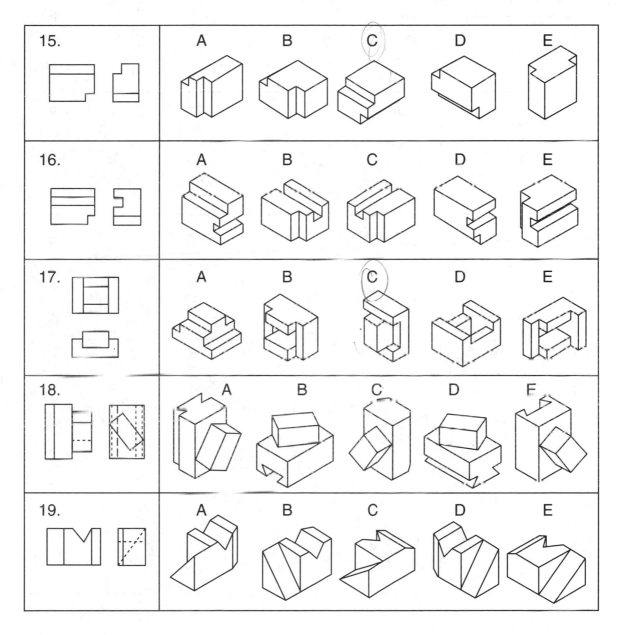

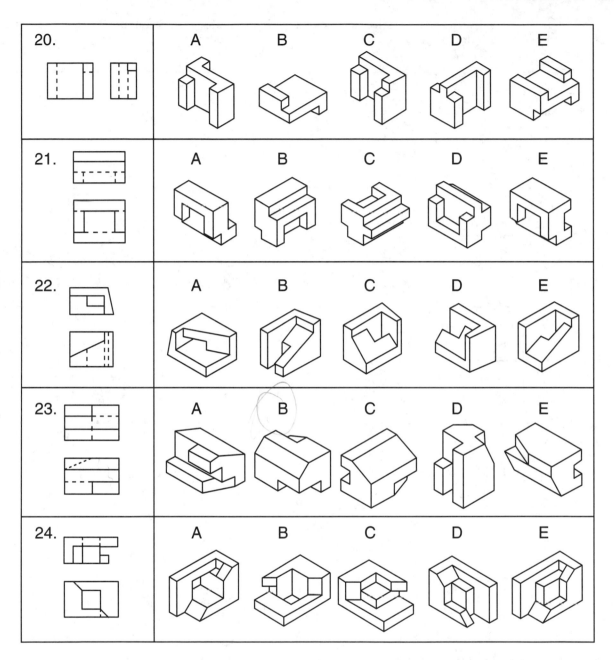

25. In the problems shown below, the circles indicate the location of missing views. Select the correct view from the 30 proposed views shown. A view may be used more than one time.

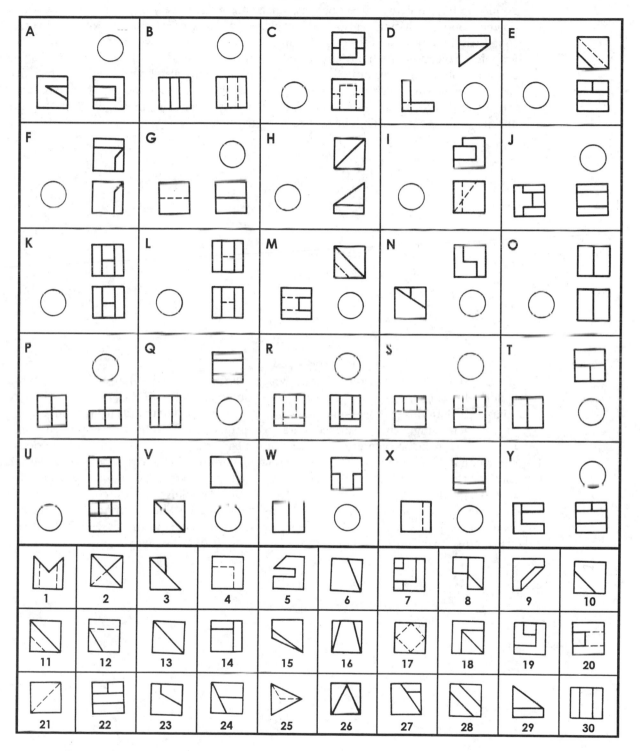

5.4 SINGLE-CURVED SURFACES

So far, we have only examined objects which consisted entirely of planar surfaces. Many objects you will encounter also contain surfaces which are curved in space. For this reason, it will be important for you to be able to communicate curved surfaces in an orthographic projection system. Single-curved surfaces are defined as surfaces having curvature about one axis only. Single-curved surfaces may also be referred to as cylindrical surfaces. A single-curved surface is generated by revolving a line about an axis. If the line is parallel to the axis, a cylinder is created; if the line is inclined with respect to the axis, a cone is created. This is illustrated in Figure 5-32.

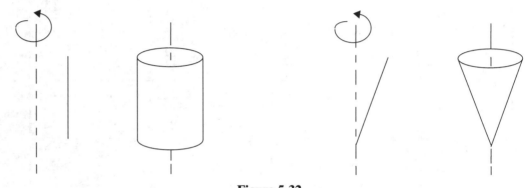

Figure 5-32

A double-curved surface is one which has curvature about two axes, making it spherical in shape. Examples of objects with double-curved surfaces would be a globe or a basketball. In this text, we will limit our discussion to single-curved surfaces only.

When a single-curved surface is drawn in profile (along a line of sight perpendicular to the axis of the surface), the surface will appear as an area. In the case of the object shown on the left in Figure 5-32, this is a rectangular area, in the case of the object shown on the right, it is a triangular area. The area is not bounded by edges, but rather by the visible extents of the surface. You can readily see this by holding up a cylindrical object such as a can and viewing it from the side. This is the one instance in engineering drawing where an object line is not used to represent an actual edge. In this case, the object line represents the cylindrical boundary, or visible extents of the surface.

When single-curved surfaces are used in engineering drawings, centerlines are usually included. The centerline indicates the radial center of the surface and is drawn along its longitudinal axis. This is the point from which the radius of curvature may be measured. The format for a centerline is a long segment followed by a short dash followed by another long segment. This format may be repeated if the feature dimensions are large. One centerline is shown in each of the area views of the single-curved surface, and two crossing centerlines are used in the circular view to indicate the point view of the radial axis. The centerlines should extend beyond the boundary of the feature. This is illustrated in Figure 5-33.

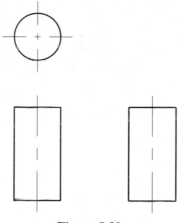

Figure 5-33

Probably the most common occurrence of single-curved surfaces in engineering drawings is as internal single-curved surfaces or holes. These internal surfaces follow the same principles as the external curved surfaces, except the cylindrical boundaries will be shown as hidden lines. Figure 5-34 shows a three-view drawing of a block with two cylindrical holes in it. One hole extends all the way through the block. This is referred to as a thru hole. The other hole does not pass through the block. This is referred to as a blind hole.

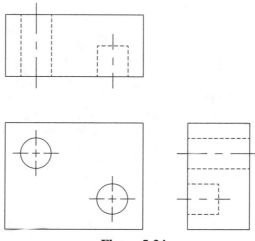

Figure 5-34

Single-curved surfaces will appear as ellipses in isometric views. An ellipse can be easily sketched using the four arc method. This involves the creation of four tangent arcs, with the opposite pairs being mirror images of one another. To sketch a cylinder using isometric grid paper, begin by drawing the visible or upper ellipse. Locate the four grid points depicted in Stage 1 of Figure 5-35. These four points lie on the ellipse, which will be the top surface of the cylinder. The diagonal distance between opposite pairs (two units in this case) corresponds to the diameter of the cylinder. Next sketch the two longer arcs as shown in Stage 2. Finish by sketching the two remaining shorter arcs to form the complete ellipse. These remaining arcs must be tangent to the first two arcs and will intersect at the radial points. This is shown in Stage 3. To finish the isometric sketch of the cylinder, show the cylindrical boundary of the cylinder with straight lines tangent to the smaller arcs. These points correspond to the endpoints of the major axis of the ellipse. Once the straight edges are drawn, sketch a half ellipse at the bottom of the cylinder. These steps are depicted in Stage 4 of Figure 5-35.

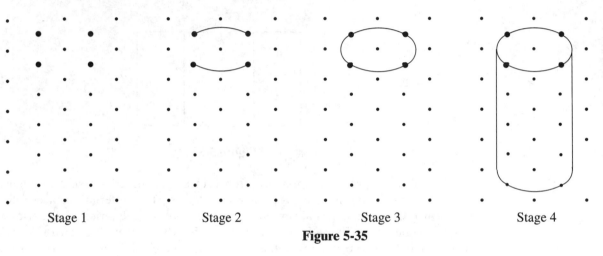

Stage 1   Stage 2   Stage 3   Stage 4

Figure 5-35

Notice that the sketched cylinder corresponds to the concepts of surfaces and edges. The cylinder consists of three surfaces (two planar and one cylindrical) which intersect in two circular edges. If we modify our original cylinder by creating new surfaces, we will note that these concepts will be maintained. Examine the cylinder in Figure 5-36. A new feature was made by cutting into the cylinder along a plane parallel to the longitudinal axis and then perpendicular to that axis. The result of these cuts are the creation of two normal surfaces on the object. Figure 5-37 shows the orthographic views of this same object. You can see that for both normal surfaces the representation characteristics established in Section 5-2 still hold true. Each normal surface appears as a true size area in one view and as a line in the other two.

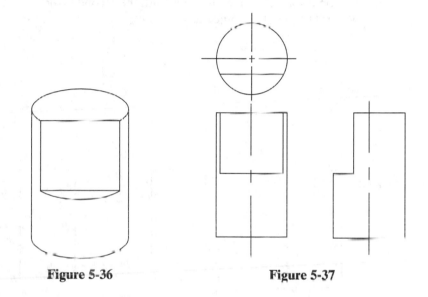

Figure 5-36   **Figure 5-37**

To add this feature to our isometric sketch of Figure 5-35, we will create an edge between the two points on the upper circular curve which establish the location of the cut feature. Two straight lines are added from the endpoints of this line to indicate the "depth" of the cut along the longitudinal axis. A line is added between the two open endpoints and an arc created between these same two endpoints to represent the intersection of the cylinder and the cut perpendicular to the cylinder axis. This construction is shown in Figure 5-38.

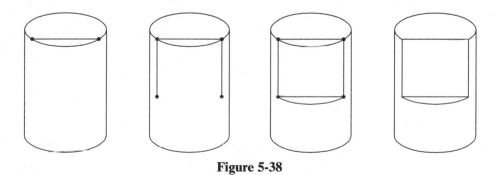

Figure 5-38

Engineering drawings are typically set up in the familiar L-shaped pattern which shows the front, top, and right side views. In this case, the side view is projected from the front view and the top view is also projected from the front view, hence the L-shape. Sometimes, however, the same three views will be shown except that both the front and the right side views are projected from the top view. If we go back to the concept of the object being surrounded by a glass cube, for this type of projection we are folding up the front and right panes of glass so that they are lying in the same plane as the top view. This type of projection is generally used for cylindrical parts where the circular view is shown in the top. In this way, the "rectangular" view of the cylinder is always projected from the "circular" view for clarity's sake. Figure 5-39 shows a comparison of the two types of projection for a cylinder. Notice that with the second projection it is, indeed, clearer what the general shape of the object is.

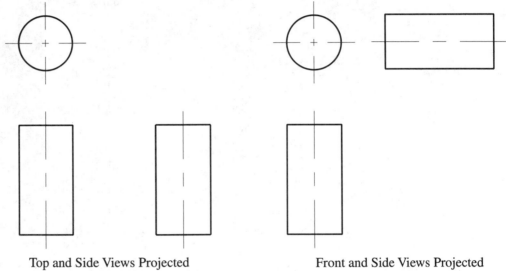

Top and Side Views Projected
from the Front View

Front and Side Views Projected
from the Top View

Figure 5-39

EXERCISES 5.4

In Exercises 1–3 sketch the top, front, and right side views for the objects shown below. Make sure you include centerlines in your sketches.

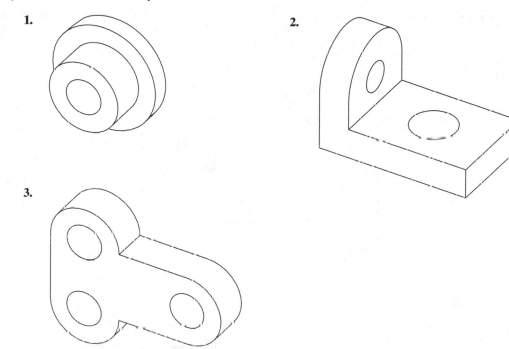

1.

2.

3.

For Exercises 4–6, the top and front views of an object are given. Copy the two views onto square grid paper and then construct the right side view for each. On a separate sheet of isometric grid paper, construct the isometric view of the object.

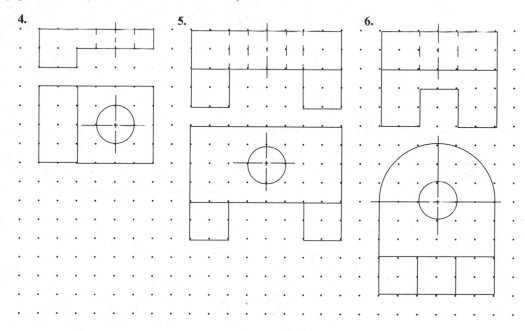

4.

5.

6.

7. In the problems shown below, the circles indicate the location of missing views. Select the correct view from the 30 proposed views shown (Note that all of the problems contain single-curved surfaces).

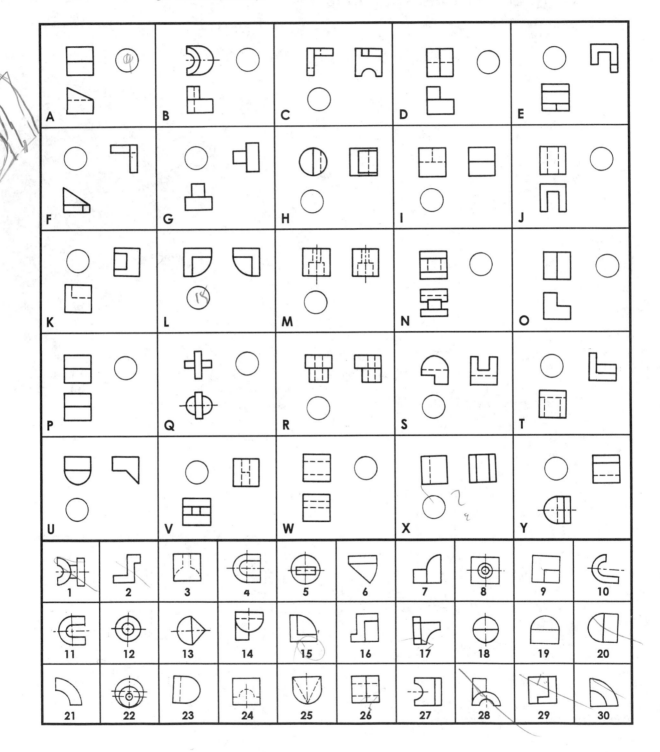

8. In the problems shown below, the circles indicate the location of missing views. Select the correct view from the 30 proposed views shown (Note that all of the problems contain single-curved surfaces).

9. In the problems shown below, for each object shown in pictorial view, select one of the 20 views that correctly shows a top, front, or right side view of the object having one or more hidden lines (Note that not all of the problems contain single-curved surfaces).

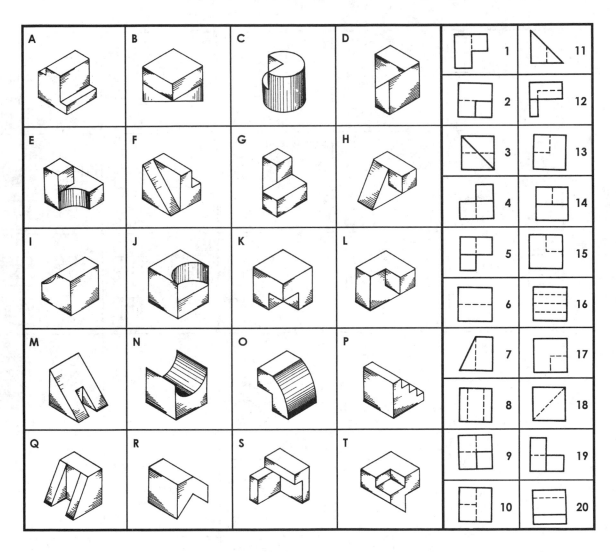

C H A P T E R

Drawing Standards and Conventions

Engineers use drawings to convey their design ideas and to communicate with one another. It is important that you learn about the conventional practices in engineering drawings in order to effectively communicate by this means. Through the years, various standards and conventions have developed in the field of engineering design representation. These standards and conventions ensure that drawings are clearly understood by all parties involved. This chapter will describe some of the standards and conventions you may find on engineering drawings. The ideas presented in this chapter will not comprise a complete list of all of the possible standards and conventions but will highlight some of the most common ones. As you work as a practicing engineer, you will learn about the standards and conventions which are relevant to your particular company and/or industry. This chapter will focus on some common aspects of conventional practice as they relate to engineering drawings.

6.1 SCALES

In Chapter 4, you learned about the object transformation of scaling. If an object is scaled, this means that the true size of the object is different from the true size of its image. Figure 6-1 illustrates this use of the word *scale*. From the lengths of the line

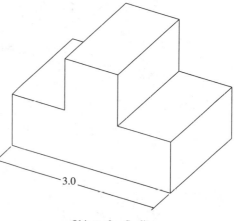

Figure 6-1   Original Object   Object after Scaling

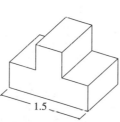

segments, you can tell that the object at the left has been dilated (or scaled) by a factor of 2 to produce the image shown on the right.

A more prevalent use of the term *scale* in engineering is when the object is drawn so that it does not appear in its true size on paper. This is known as drawing an object to scale. By drawing objects to scale, engineers, architects, and city planners are able to communicate their designs to one another and to the people who implement these designs. Imagine how difficult it would be to design a building effectively if the architect were only able to draw it at its true size. A piece of paper would have to measure 50 feet × 100 feet or more. You are probably familiar with the concept of drawing something to scale. If you have ever looked at a map, a scale for the map is usually included in the legend and you use that scale to help you determine the actual distance between two points on the map.

Figure 6-2 illustrates the concept of drawing something to scale. Here the objects are the same size—the one on the right just *appears* to be twice as large as the one on the left. The notation of a 2:1 scale means that two inches on paper equal one inch on the object. The drawing therefore looks larger than the actual object. Conversely, a 1:2 scale would mean that one inch on paper represents two inches on the physical model; hence, the drawing would look smaller than the actual object. In denoting scales, the first number in the ratio corresponds to the drawing, and the second number in the ratio corresponds to the physical object. A similar principle applies to scaled models.

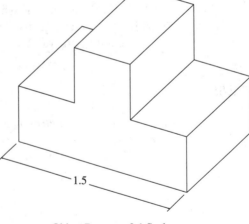
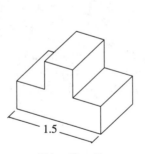

Object True Size   Object Drawn at 2:1 Scale

Figure 6-2

Another way to think of drawing objects to scale is that the scale merely indicates how close you are to the object. If you are a substantial distance from the object, it appears very small; whereas, if you are very close to the object, it appears large. However, the true object size does not change. Similarly, from an airplane, a house on the ground looks tiny, but if you are standing a few inches from the same house, it appears enormous. Clearly, it is the perception of the house that changes, not the size of the house. The concept of drawing an object to scale has previously been illustrated in Chapter 2 of this text in our discussion of surveying traverses. Figure 2-21 shows a traverse that, in reality, measures hundreds of feet but on the page measures only a few inches.

Scales for drawings are usually reported as ratios. However, sometimes drawing scales are denoted with an equal sign rather than a ratio. This is particularly true when scales are given in the English system of units. Thus, a scale may be reported as $1'' = 50'$ or $1/4'' = 1'-0''$. In the first case, this means that one inch on the drawing corresponds to 50 feet on the actual object. These drawing scales can also be related back to their ratio equivalents, i.e., $1'' = 50'$ corresponds to a scale of 1:600 (there are 12 inches in a foot, so $50' = 12 * 50 = 600''$, thus, the 1:600 size ratio), and a scale of $1/4'' = 1'-0''$ corresponds to 1:48 (a ratio of $1/4''$ to $12''$ is the same as $1''$ to $48''$).

When drawing or sketching something to scale, many novices will use a calculator to accomplish this. For example, if you were to draw something at a scale of $1'' = 40'$ and you needed to draw a line which represented 25' on the object, you could calculate that the line you should draw will be 25/40 or 5/8'' long. Similarly, if you needed to draw a line which represented 80' on the object, you could draw it 80/40 or 2'' long. Using a calculator to figure out how to make the drawing would be extremely tedious and time consuming. Fortunately, this type of tedium is not necessary in the preparation of engineering drawings or sketches. A scale is a device which is used by drafters and engineers to aid them in making a drawing to scale or in reading a dimension from a scaled drawing. A scale is typically a triangular prism in shape, and thus there are usually six to twelve different drawing scales depicted on one piece of equipment. There are three predominant types of scales which are currently available on the market: an Engineer's scale, an Architect's scale, and a Metric scale.

Engineer's Scale. An Engineer's scale (sometimes called a Civil Engineer's scale) usually consists of 10, 20, 30, 40, 50, and 60 scales. These scales are based on the English system of units, with the inch as the basis for measurement. The divisions are in increments of tenths of an inch (not eighths as you are probably familiar with on ordinary rulers). Engineer's scales can be used to draw in any multiple of ten of the basic units. For example, the 30 scale can be used for making or reading the following scales from drawings: $1'' = 3'$, $1'' = 300'$, or $1'' = 30$ mi. Similarly, the 50 scale can be used for drawings with scales of $1'' = 50'$, $1'' = 5$ yds, or $1'' = 500$ mi.

Figure 6-3 shows a line being measured by use of a 20 scale. The actual length of the line on paper is 3.5″, but since it is drawn at a scale of 1″ = 200′, this line represents a length on the actual object of 700′. By reading the scale in this figure, what is the length of the actual line to point A? As can be seen in this figure, the actual line to point A is 580′ long. Measure this line using a regular inch ruler and verify that the length of the line to point A, as it is drawn on the paper, is equal to 2.9″.

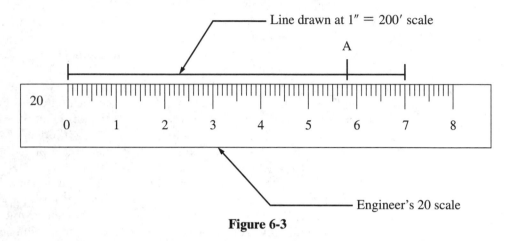

Figure 6-3

Figure 6-4 shows lines drawn at a scale of 1″ = 40′ and a corresponding 40 Engineer's scale. What is the length of each line segment (OA, OB, OC)? These values can be read directly from the scale, making the proper adjustments for decimal places. The line is drawn at a scale of 1″ = 40′; therefore, the first "2" on the scale represents 20′, the "4" represents 40′, the "6" represents 60′, and so on. The unlabeled long tic marks represent 10′, 30′, 50′ and so on. The intermediate-length tic marks occur at 5′ intervals, and the smaller tic marks represent 1′ intervals. Therefore, the length of the line OA can be read as 67′ on this scale. The line OB is read as 103′. Many times, novice scale readers will incorrectly interpret the length of OB as 130′. When reading the scale, care should be taken to put the decimal in the correct place. The distance OB is read directly as 10.3, but since the scale is 1″ = 40′, the decimal is moved to the right one unit and the correct scale reading becomes 103′. The length of the line OC can be read from the scale as 132′.

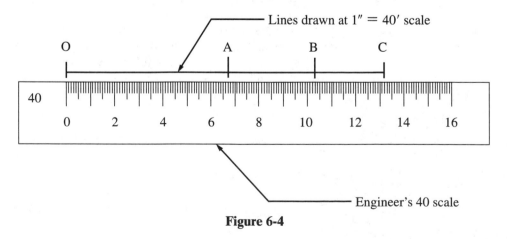

Figure 6-4

For the lines shown in Figure 6-4, what are their lengths if they were drawn at a scale of 1″ = 4000′? Note that you still use the same Engineer's scale (the 40 scale) to make this reading, but that you add more zeros to the number that you read on the scale. Thus, the line OA has a length of 6700′, OB has a length of 10,300′, and OC has a length of 13,400′ at a scale of 1″ = 4000′. What if the lines were drawn at a scale of 1″ = 4 yds? At this scale, the line OA has a length of 6.7 yds, OB has a length of 10.3 yds, and OC has a length of 13.4 yds.

Architect's Scale.   An Architect's scale is similar to an Engineer's scale in that it is based on the English system of units. One of the differences between the two scales is that the Architect's scale is based on fractions of an inch (recall that the Engineer's scale is based on tenths). Another significant difference is that with an Architect's scale, drawing scales are always reported as: something = 1′-0″. Thus, a scale might be reported as 1/4″ = 1′-0″ or as 3/8″ = 1′-0″. Some of the more common scales which are depicted on an Architect's scale are listed in Figure 6-5.

12″ = 1′-0″ (full size)	1/2″ = 1′-0″ (1/24 size)
6″ = 1′-0″ (half size)	3/8″ = 1′-0″ (1/32 size)
3″ = 1′-0″ (1/4 size)	1/4″ = 1′-0″ (1/48 size)
1 1/2″ = 1′-0″ (1/8 size)	3/16″ = 1′-0″ (1/64 size)
1″ = 1′-0″ (1/12 size)	1/8″ = 1′-0″ (1/96 size)
3/4″ = 1′-0″ (1/16 size)	3/32″ = 1′-0″ (1/128 size)

Figure 6-5

Architect's scales will usually look significantly different than Engineer's or Metric scales. The biggest difference is that each edge of the Architect's scale typically depicts two scales— one reading from left to right and the other reading from right to left. Thus, twice as many scales (12 versus 6) are depicted on an Architect's scale when compared to an Engineer's or Metric scale. The other difference is that fractional gradations are only shown at the end of the scale. Thus, when you are measuring a distance with an Architect's scale you must place one end of the line at the nearest whole number foot on the scale and read the fractional foot in the end with the gradations. Figure 6-6 shows an Architect's 1/4 scale (this means that 1/4″ = 1′-0″) and a line to be measured at this scale.

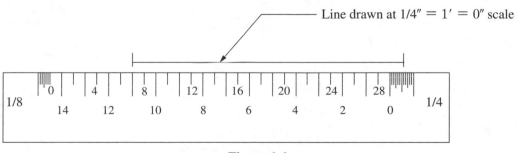

Figure 6-6

In this figure, the 1/4 scale is shown reading right to left on the scale and the 1/8 scale is shown reading from left to right. To read the length of this line, one end of the line is placed on the nearest even foot mark of the scale. In this case, it is 11′ (remember that you are reading from right to left for the 1/4 scale). Notice that the closest foot mark is not labeled for you. For the 1/4 scale, the even foot markers are labeled and the odd ones are not. The smaller tic marks for this scale represent one-half foot divisions. Be careful not to line up the end of the line with the half-foot marks instead of the foot markers. The fractional feet are shown in the last foot of the scale (past the 0). This last foot is divided into 12 gradations, so each tic mark on the scale represents 1″ because there are twelve inches in one foot. Thus, the length of this line is 11′-7″. What happens if you read the length of this line using the 1/8 scale? Figure 6-7 shows the same line being measured with the 1/8 scale. The same procedure is followed to read the length of the line at this scale. In this case, the nearest even foot mark is at 23′. Notice that at this scale only every fourth (4, 8, 12, etc.) foot marker is labeled and that the long tic marks reading left to right represent the odd foot marks. The short tic marks reading from left to right represent the intermediate even foot markers. The final foot of this scale is divided into six even increments compared to twelve increments for the 1/4 scale. Thus, each division in the final foot represents 2″ for this scale. As it can be seen in this figure, the length of the line at a 1/8 scale is 23′-2″ (note that this makes sense because two times 11′-7″ is 22′-14″ or 23′-2″). The Architect's scale will usually take a great deal of practice on your part to read it with confidence.

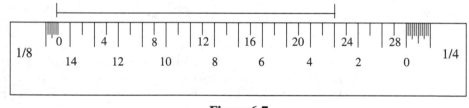

Figure 6-7

Metric Scale.   Metric scales are similar to the scales which were previously discussed except that they are based on the Metric system of units. Because metric units are based on decimals, unlike the English system of units (12″ per foot, 3′ per yard, 1760 yards per mile, etc.), Metric scales are always reported as ratios. Thus, typical Metric scales are reported as: 1:1, 1:2, 1:5, 1:10, and so on. The same principal which is used to measure distances with either an Architect's or Engineer's scale is used for Metric scales. Like the Engineer's scale, the Metric scales can be used for multiples of ten of the basic unit. Thus, a 1:5 scale can also be used to measure 1:50, 1:500, and 1:5000 scales. Figure 6-8 shows a line drawn at a 1:2000 scale and a 1:20 metric scale. On a Metric scale the numbers (0.5, 1.0, 1.5, etc.) generally represent meters and you adjust the decimal according to the specific scale at which you are measuring. If you read the scale directly, the length of the line is 1.84 m. But since the scale depicts a 1:20 drawing scale and the line was drawn at a scale of 1:2000, you must move the decimal place two units to the right to account for this difference. Thus, the length of the line is equal to 184 m. Similarly, if the line were drawn at a 1:20 scale, the length of the line would be read as 1.84 m, or if it were drawn at a 1:200 scale, its length would be 18.4 m.

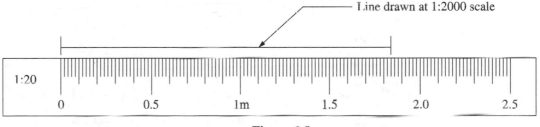

Figure 6-8

Figure 6-9 shows a Metric scale of 1:100 and a set of lines drawn at a 1:1 scale. What are the lengths of each line? If you read the length of OA directly from the scale, you see that it is 7.7m. Since the scale is 1:100 and the line is drawn at 1:1, you move the decimal place two units to the *left*, therefore the length of the line is 0.077 m. Alternatively, you could report the length of the line as 7.7 cm or 77 mm. What is the length of OB? The value read from the scale is 10.7 m. If you move the decimal place two units to the left, the length is determined as 0.107 m (*or* 10.7 cm *or* 107 mm). Similarly, the length of line OC is 0.124 m (12.4 cm).

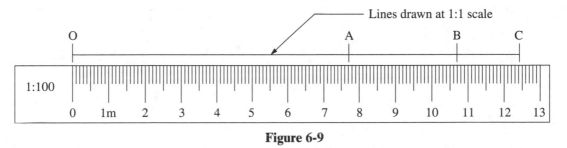

Figure 6-9

As you hopefully can see from the previous discussion of drawing scales (Engineer's, Architect's, and Metric), these devices are time saving and relatively easy to use. Unfortunately, many students have a tendency to try to use their calculators when working with scaled drawings for the first time. It is important that you learn to properly use a scale for reading or making engineering drawings.

EXERCISES 6.1

For the following exercises, use the indicated scale to *measure* the lengths of the lines, i.e., do <u>not</u> use a calculator for these exercises.

For the figure shown below, what are the lengths of the lines A through F if they are drawn at the indicated scales.

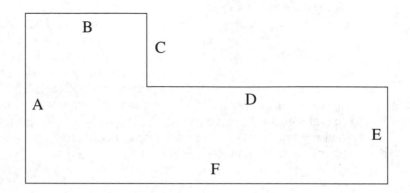

1. 1″ = 4000′
2. 1″ = 5 yds
3. 1″ = 60′
4. 1/4″ = 1′-0″
5. 3/8″ = 1′-0″
6. 3/4″ = 1′-0″
7. 1/2″ = 1′-0″
8. 1:2 (use a Metric scale)
9. 1:500 (use a Metric scale)
10. 1:75 (use a Metric scale)

6.2 *CONVENTIONAL PRACTICE AND ORTHOGRAPHIC PROJECTION*

There are many instances when the strict rules of orthographic projection are not followed according to conventional practice. One case where this is true occurs when two surfaces intersect in a smooth curve rather than in a sharply defined edge. In this case, following the rules of orthographic projection would mean that object lines would not be drawn at the intersection of the surfaces. However, leaving out the line of intersection between the two surfaces would likely lead to confusion about the appearance of the object. In this case the rules of orthographic projection are violated in favor of conventional practice, which dictates that the implied line of intersection between the two surfaces be drawn. Figure 6-10 shows an example of this convention. The left side view of the object shows a cylindrical surface intersecting a flat surface in a smooth transition. By the rules of orthographic projection, an object line should not be drawn in the front view of the object (labeled "True Orthographic Projection" in Figure 6-10). However, without the object lines, the drawing of the object is somewhat confusing about its exact appearance. Conventional practice dictates that object lines be drawn at the implied intersection for clarity when the difference in the sizes of the two arc's radii are sufficiently large. Figure 6-11 shows an example of an object which contains arcs that are nearly the same size. Note that the true projection is preferred for this object.

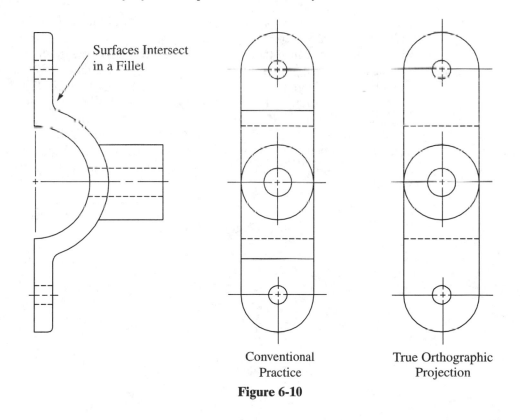

Surfaces Intersect
in a Fillet

Conventional
Practice

True Orthographic
Projection

Figure 6-10

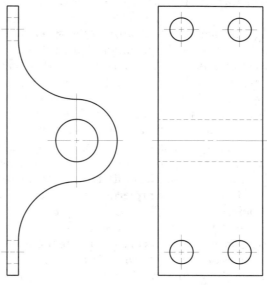

Figure 6-11

Another instance when the rules of orthographic projection are not followed in favor of conventional practice occurs when an object contains features which are located in a radial pattern such that true projection does not show the feature in its true size. In this case, the radial part of the object is revolved to an aligned position when constructing the adjacent orthographic view. An example of this is shown in Figure 6-12. In this case, the three small holes in the object are located radially from the center of the object 120° apart as seen in the front view. Conventional practice dictates that the two "upper" holes be revolved to their vertical radial position as shown in the right side view labeled "Conventional Practice." Also shown in this figure is the true projection of the holes. Note that the true orthographic projection is confusing and does not convey the true object geometry well.

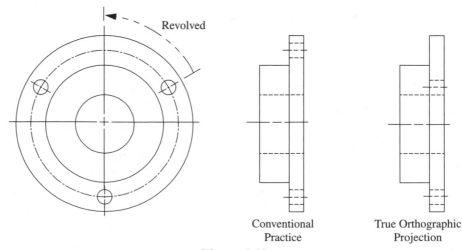

Figure 6-12

Another example of an object with a feature revolved so that it is aligned with an adjacent view is shown in Figure 6-13. In this figure, the object as shown in the front view has an angled feature on its right side. The top view has been drawn according to conventional practice. In the top view, the angled part of the object has been revolved so that it is aligned with the rest of the object. Aligning the view in this way better depicts the significant features of the object and shows the holes at their true distance from one another.

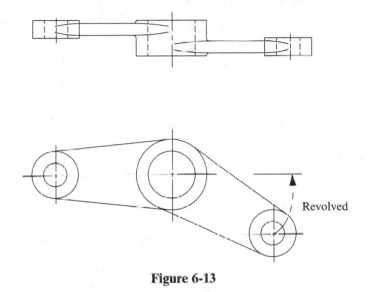

Revolved

Figure 6-13

Another instance where the rules of orthographic projection are violated is in depicting intersections between cylindrical parts where the cylinders have a large difference in their diameters. When a feature on an object intersects a cylindrical part, a curved line of intersection results. This curved line is generally depicted as a straight line by conventional practice. Figure 6-14 shows an example of this situation. The object shown in this figure is a hollow cylindrical shape with a cylindrical "post" protruding from the bottom and a cylindrical hole through it on the top. Because the walls of the tube are curved,

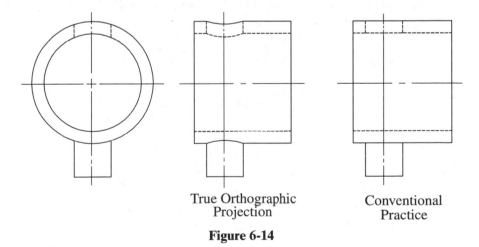

True Orthographic
Projection

Conventional
Practice

Figure 6-14

the intersections between the tube and the post, and the tube and the hole form curved lines. Conventional practice dictates that these intersections be portrayed on the drawing as straight lines, as shown in this figure (labeled "Conventional Practice"). If, however, the curvature of the intersection were significant, as when the two diameters are closer in size, then you might be required to show the intersection in true projection. Figure 6-15 shows the conventional practice which is followed when the intersecting cylinders have the same or similar diameter.

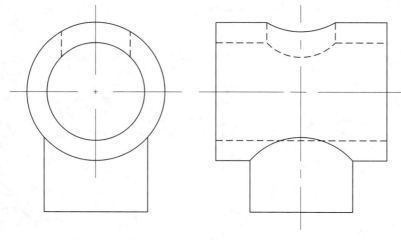

Figure 6-15

Many of the drawing conventions described in this section will not be easily achieved with computer modeling software. For example, suppose you create a 3-D solid model of a part which contains a curved intersection between two cylindrical parts. If you then create a drawing from the solid part, the intersection will be shown as a curved line in the drawing since the computer is working from the actual model database. For this reason, many of the conventions described in this section may change as 3-D computer modeling becomes more prevalent and the accepted practice. If you are sketching objects, however, you will not have any problems in meeting these conventions.

EXERCISES 6.2

In Exercises 1–5, sketch two views of the indicated objects according to conventional practice.

1.   **2.**   **3.**

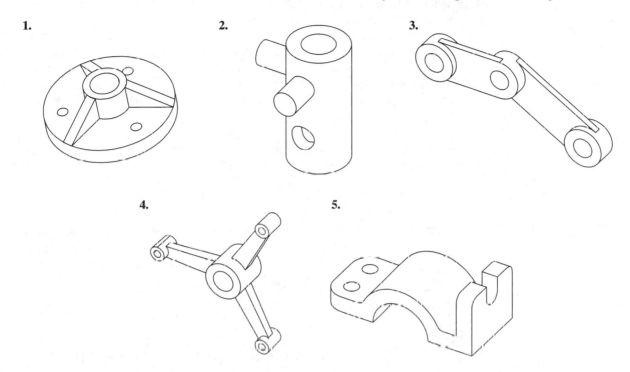

4.   **5.**

6.3   *VIEW SELECTION*

When constructing engineering drawings or sketches, you should be careful to include the minimum number of views which are necessary to define all of the features of the object. In general, this means that you will have either a three, two or one principal-view drawing plus section and auxiliaries as required (sections and auxiliaries will be covered in subsequent chapters). When selecting the number of views which are required, you will typically select the most descriptive view as the front view. The reason for this is that when you add dimensions to the drawing, it is best to have most of the dimensions around the front view. Good dimensioning practice dictates that dimensions be included in the view which best describes the outline of the features of the object, therefore it is desirable to have the most descriptive view in the front. You will learn more about dimensioning practice in Chapter 9.

Although two views specify all three spatial dimensions, many objects will require three views to adequately show all of their features. More complicated objects may even require isometric pictorial, auxiliary views (Chapter 7) or section views (Chapter 8) for clarity's sake. Fortunately in this era of 3-D computer modeling software, it is relatively easy to include an isometric pictorial as part of the final drawing. If three views are required to describe an object, it is customary to include the top, front, and right side views as the standard orthographic views. Sometimes, the left side view is chosen instead of the right side view in a drawing layout. The rule of thumb you should follow is that you choose the view to minimize the number of hidden lines on the view. Figure 6-16 shows a three-view drawing of an object. Notice that all three views are required in order to convey a clear idea of the various features found on the object. Note also that if the left side view were selected instead of the right side view, more of the features would be hidden from this viewpoint; therefore, it is better to include the right side view for this particular object. For this object, its basic Z-shape is apparent in the top view, the front view shows the cut-out feature the best, and the inclined surface as well as the projection towards the back of the object is shown best in the right side view. Try to imagine understanding what this object looks like if you only had two views displayed (it would be difficult!).

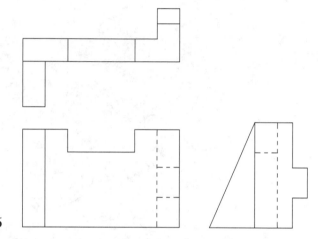

Figure 6-16

Some objects can be adequately described with only two views. Typically the two views which are chosen are either the top and front views or the right side and front views (you can choose the left side view in place of the right side view if it improves clarity). Again, the most descriptive view should be placed in the front view for dimensioning purposes. Figure 6-17 shows a two-view drawing of an object. Notice that all of the features of the object are described with just these two views. If the right or left side views were included, no additional information regarding the features of the object would be gained. The basic U-shape of the object is shown in the front view, and the location of the hole and the size of the notch in the object are visible in the top view. Thus, the side views would only add extraneous information which should be avoided.

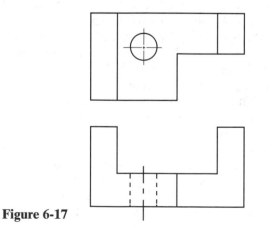

Figure 6-17

Certain objects can be completely described with only one orthographic view. The type of object which can be described with one view is either a flat object with a constant thickness or a cylindrically shaped object such as a shaft, a bolt or a screw, or a spherical object. A note is typically included on a single-view drawing that indicates the thickness or diameter of the part. Figure 6-18 shows some examples of objects which are defined by single-view drawings.

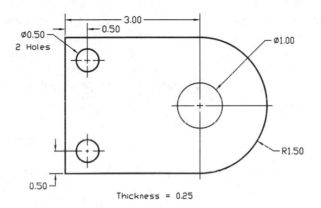

Object of Constant Thickness

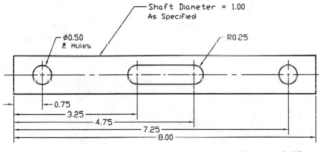

Cylindrical Object of Constant Diameter

Figure 6-18

Engineering drawings are typically set up in the familiar L-shaped pattern which shows the front, top, and right side views. In this case, the side view is projected horizontally from the front view and the top view is projected vertically from the front view, hence the L-shape. Sometimes, however, the same three views will be shown except that both the front and right side views are projected from the top view. Still other times, the top, front, and left side views are shown for an object. If we go back to the concept of the object being surrounded by a glass cube, these different drawing layouts are obtained by the manner that the panes of glass were "unfolded." For a drawing which shows the top, front, and left side views, the top and left panes of glass are unfolded to lie in the same plane as the front view. For a layout which shows the front and right side views projected from the top, we are folding up the front and right panes of glass so that they are lying in the same plane as the top view. These drawing layouts are illustrated in Figure 6-19.

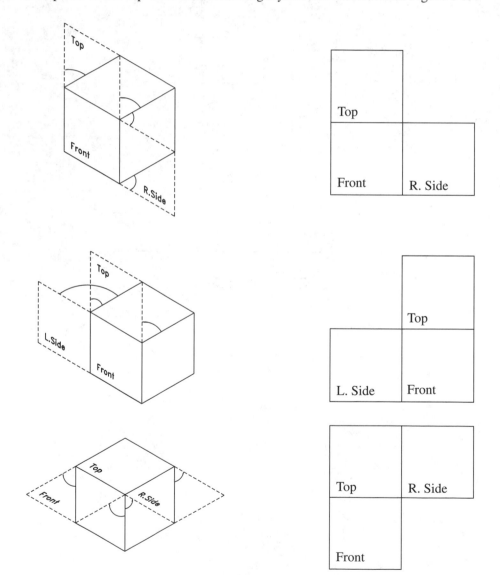

Figure 6-19

Drawing layouts which show the front and right views projected from the top are generally used for cylindrical parts where the circular view is shown in the top, as discussed previously in Chapter 5. In this way, the "rectangular" view of the cylinder is always projected from the "circular" view for clarity's sake. Figure 6-20 shows a comparison of the two types of projection for a cylinder. Notice that with the second projection it is, indeed, clearer what the general shape of the object is.

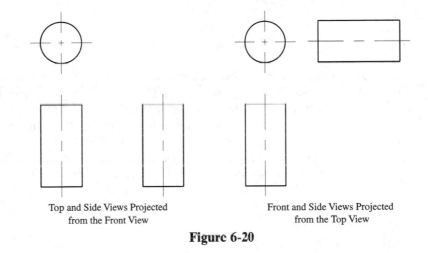

Top and Side Views Projected
from the Front View

Front and Side Views Projected
from the Top View

Figure 6-20

EXERCISES 6.3

In Exercises 1 5, sketch the required orthographic views to completely describe the object.

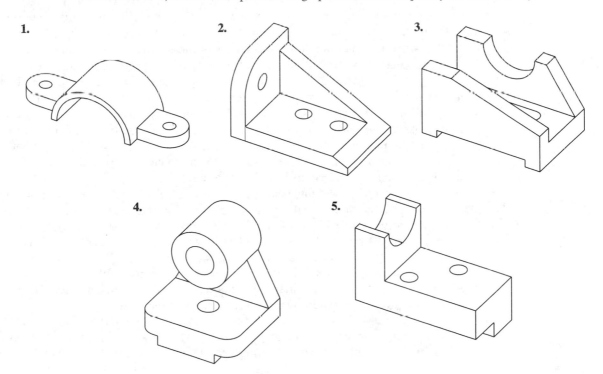

1.

2.

3.

4.

5.

C H A P T E R 7

Auxiliary Views

• •

As you learned in Chapter 5, inclined and oblique surfaces are not seen in true size in any of the principal orthographic views. The reason for this is that these types of surfaces are not parallel to any of the principal viewing planes. Many times when you are constructing an engineering drawing or sketch you will need to show these surfaces in true size. To do this, you will need to construct a view which is parallel to the surface in question. In this chapter you will learn how to construct such a view for inclined surfaces. Creating a true-size view of an oblique surface requires additional view constructions and will be covered in Section 7.3.

7.1 CREATING AUXILIARY VIEWS

Any view of an object which is not one of the principal views is called an auxiliary view. Thus, if you construct a new view of an object which shows an inclined surface in true size, this will be an auxiliary view. Typically when creating an auxiliary view, you will only be interested in viewing the inclined surface true size and will not be interested in viewing the entire object from this vantage point. Thus, in this section we will be primarily interested with constructing the auxiliary view of the inclined surface only.

To understand how auxiliary views are created, it is useful again to think about the object as if it were surrounded by a glass cube. Imagine that an extra pane of glass were somehow inserted in the glass box and the orientation of this extra pane of glass was parallel to the inclined surface of the object. If the inclined surface is projected onto this pane of glass in the same way that surfaces are projected onto one of the six principal planes, then the projected view of the surface will be true size because the pane of glass is parallel to the surface. Figure 7-1 shows an object which is surrounded by a glass cube, with only the inclined surface projected onto a parallel pane.

If you then unfold the panes of glass from this cube so that they all lie within one plane, and rotate the plane until it is in the plane of the paper, you will end up with the drawing shown in Figure 7-2. The auxiliary view now shows the inclined surface in true size. Notice that the size of the surface in this view is larger than its size in either the side or top views. This is due to the fact that the principal views of the surface are foreshortened as described in Chapter 5. Notice also that when the glass cube is unfolded, the fold line for the auxiliary view is parallel to the line view of the inclined surface. The line view of an inclined surface is sometimes referred to as the edge view.

The question is: How do you construct this auxiliary view of the inclined surface on a 2-D sheet of paper, without the benefit of a glass cube? In Figure 7-2 the fold lines where the glass cube was "unfolded" are shown. Recall that the fold line for the auxiliary view is

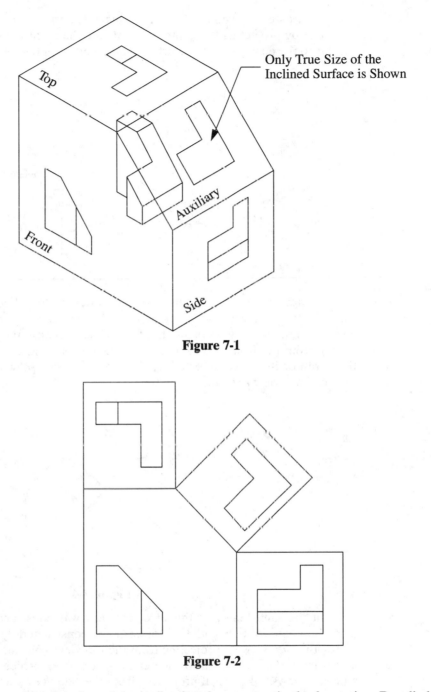

Only True Size of the
Inclined Surface is Shown

Figure 7-1

Figure 7-2

parallel to the edge view of the inclined surface as seen in the front view. Recall also that when projecting from one view to another, projection rays which are perpendicular to the fold lines are used in a system of orthographic projection. Thus, the projection rays from the front view into the auxiliary view will be perpendicular to the fold line between the two views. This also means that the projection rays are perpendicular to the line view of the inclined surface.

To create an auxiliary view of the inclined surface which shows it in true size, you first start by projecting the points which define the surface along rays which are perpendicular to the edge view of the surface. This is shown in Figure 7-3.

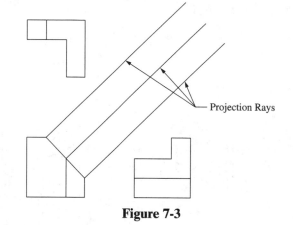

— Projection Rays

Figure 7-3

To draw the surface in this auxiliary view, you should note that any view projected from the front view will show the depth of the object along the projection rays. The side and top views also show the depth of the object because they were constructed by projecting from the front view along perpendicular projection rays. Thus, you can measure the depth of the object from either the top or the side view and transfer the depth into the auxiliary view as shown in Figure 7-4.

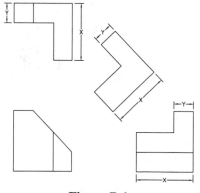

Figure 7-4

In the same way that this auxiliary view was constructed by projecting from the front, an auxiliary view could just as easily be constructed by projecting from either the top or the side view. In projecting from the top view, you should note that any view projected from the top will show the height of the object which is visible in both the front and side views. Likewise, if projecting from the side view, the width of the object will be shown which is visible in both the front and the top views. To summarize, the procedure used to create an auxiliary view which shows an inclined surface in true size is as follows: First, identify the edge or line view of the surface to be projected. Second, project the points which define the surface along rays which are perpendicular to the line view of the surface. Finally, obtain the projected dimension of the surface by observing the same dimension in an adjacent view (another way to think of this is that the dimension is found in a view which is two views away from the auxiliary view).

EXERCISES 7.1

For Exercises 1–6 shown in the following, transfer the two views shown to a sheet of plain paper (you may wish to place grid paper underneath the paper to help you transfer the views). Construct an auxiliary view of surface A which will show the surface in true size.

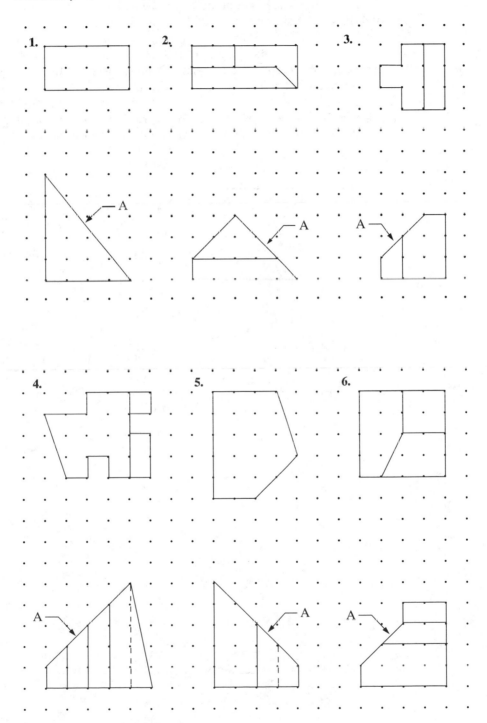

7.2 AUXILIARY VIEWS OF CURVED OR IRREGULAR SURFACES

Many times you will need to construct an auxiliary view of a surface on an object which shows a curved or irregular feature. In particular, if an object contains a hole in an inclined surface, an auxiliary view will be necessary to accurately locate the hole for machining purposes. Figure 7-5 shows a standard four-view drawing of an object which contains two holes on an inclined surface. Notice that the edge view of the inclined surface is shown in the top view, and the circular holes appear as ellipses in both the front and right side views. The procedure for creating the true-size auxiliary view of this inclined surface is essentially the same as outlined in Section 7.1. The difference is that you must first locate several points on the curved surface to project into the auxiliary view so that it is well defined. For a circular hole, usually four radial points are sufficient, but for an irregular curve, you may need to locate several points to obtain an accurate projection.

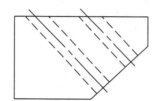
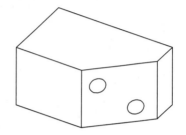
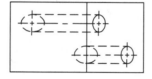
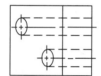

Figure 7-5

Figure 7-6 shows the first step in the construction of the true-size view of the inclined surface found on the object shown in Figure 7-5 (the centerlines have been omitted from the next two figures for clarity's sake). Note that the projection rays are constructed perpendicular to the edge view of the inclined surface. The points which are projected into the auxiliary view include the four corner points of the surface (points A, B, C, and D) as well as four radial points for each circle. Only the points defining one of the circles are labeled (point E, F, G, and H). Note also that only the top and front views of the object are shown (only two views are necessary for creating an auxiliary view).

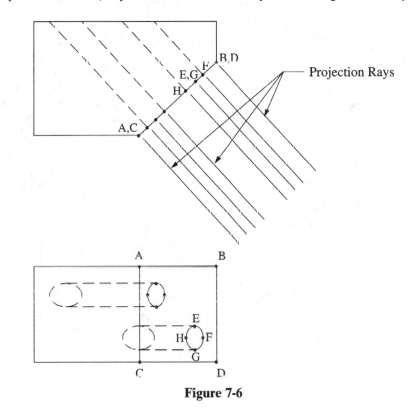

Figure 7-6

Figure 7-7 shows the distances which are transferred from the front view (two views away from the auxiliary view) into the auxiliary view. Three of the distances (a, b, and c) are necessary to define the four radial points on the circle. The distances which define the points on the other circle are not shown. The distance "x" is the height of the object as seen in the front view. Note that the top surface of the object, represented by the upper horizontal line in the front view, was used as the "fold line" in transferring the distances into the auxiliary view.

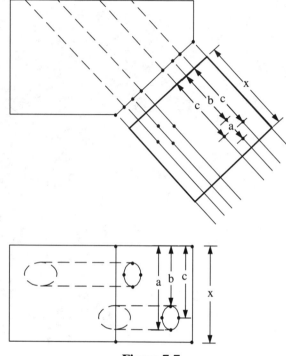

Figure 7-7

Figure 7-8 shows the completed auxiliary view for this inclined surface. Centerlines can be added to this auxiliary view and dimensions included if necessary. Dimensioning technique for locating holes is described in Chapter 9.

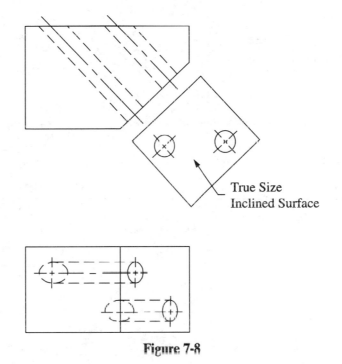

True Size
Inclined Surface

Figure 7-8

The procedure you would use to construct an auxiliary view of an irregular curved surface would be the same as outlined here. The only difference is the number of points you may have to transfer into the auxiliary view in order to define the irregular surface.

EXERCISES 7.2

Copy the two views of the objects shown in Exercises 1–4 onto a clean sheet of paper (you will probably want to "spread" the views out more) and construct the indicated auxiliary views.

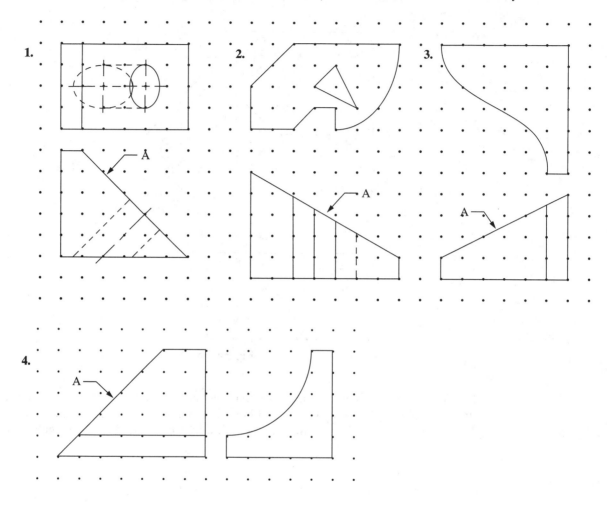

7.3 APPLICATION

Descriptive Geometry

The field of descriptive geometry was developed by French mathematician Gaspard Mongé in the mid-18th century. With descriptive geometry, engineering problems are solved by graphical methods. Three-dimensional properties and descriptions of geometric entities are used to solve complex, spatial problems. This text provides a short introduction to some of the basic concepts from descriptive geometry. For a more detailed presentation you should consult any text on the subject. You may find that some of the terminology used in this section differs from that in previous chapters. Most notably, the term "edge view" for a surface will be used in place of the term "line view" you learned in Chapter 5. This is in keeping with traditional descriptive geometry theory.

In order to work descriptive geometry problems effectively, you will not be able to free-hand sketch the solutions to the problems as you did in the previous chapters of this text. To work the problems in this section, you will probably need the following drawing instruments at minimum: 1) a pair of dividers for transferring distances, and 2) a set of triangles (one should be a 30–60–90 triangle and the other a 45–45–90 triangle).

Most of the problems you will work in descriptive geometry will require you to draw lines which are either parallel or perpendicular to other lines on the paper. You can use your set of triangles to easily do this. To draw a line parallel or perpendicular to another line, first align one leg of one of the triangles with the line. Place the hypotenuse of the second triangle (the base triangle) up against the hypotenuse of the first one and hold it firmly in place. Slide the first triangle up or down on the paper, making sure that its hypotenuse always remains in contact with that of the base triangle. As you slide the triangle, one of its legs will always remain parallel to the original line and the other leg will always remain perpendicular to the original line. You can then use either of these straight-edge surfaces to draw parallel or perpendicular lines. This method of using triangles to draw parallel or perpendicular lines is illustrated in Figure 7-9.

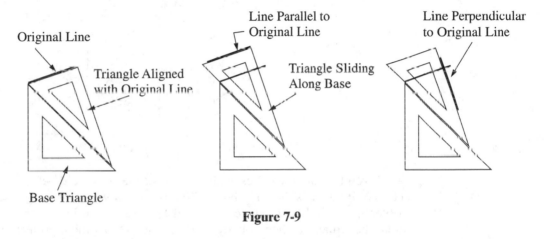

Figure 7-9

Points in Space. In Chapter 2, you learned how to plot points by their 3-D coordinate locations. In descriptive geometry you will have to locate many different points precisely in 3-D space. However, points will be described by their location relative to the principal projection planes. Descriptive geometry is based on the system of orthographic projection. In Chapter 5, you were shown how the orthographic projection system is used to describe objects in space. Here you will be concerned primarily with the location of points, lines, and planes in space. But, since lines and planes are defined by points, you must first learn how to define points within the orthographic projection system.

Recall that with orthographic projection you imagine that the object is surrounded by a glass cube and that each view of the object is projected onto the panes of glass. In this case, the "object" is merely a point in space. Figure 7-10 shows point A projected onto the panes of a glass cube which surrounds it. Note that when working descriptive geometry problems, the view which you used to call the "top" is now referred to as the Horizontal view and the "right side" view is now known as the Profile view.

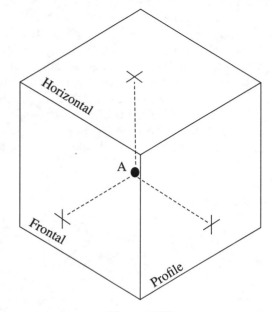

Figure 7-10

If you fold out the panes of glass for this cube as you did before, you obtain the projection of Point A as shown in Figure 7-11. Note that when working descriptive geometry problems, the *fold lines*, labeled H/F and F/P, are shown in the projection of the point unlike the situation when you are showing the orthographic projection of an object. In the front view, the H/F fold line represents the edge view of the Horizontal plane and the

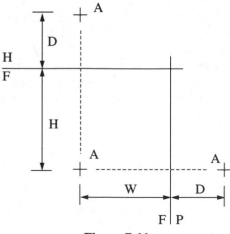

Figure 7-11

F/P fold line represents the edge view of the Profile plane. Similarly, in the horizontal view, the H/F fold line represents the edge view of the Frontal plane and in the profile view, the F/P fold line represents the edge view of the Frontal plane.

As can be seen in Figure 7-11, point A is located a distance of H below the horizontal plane, a distance of W to the left of the profile plane, and a distance of D behind the frontal plane. Note that the distance H can be measured in either the front or profile view, the distance W can be measured in either the front or horizontal view, and the distance D can be measured in either the horizontal or profile view. Thus, only two views are required in order to completely define the location of the point in space. Typically in descriptive geometry problems, only the horizontal and front views are shown. Figure 7-12 shows the Horizontal and Front views of a point B. How would you construct the Profile view of the point?

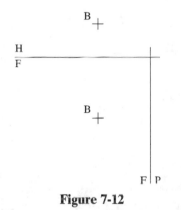

Figure 7-12

You know that, in an orthographic projection system, the points will project from one view to the next along a ray that is perpendicular to the projection plane. Thus, you can determine the projection ray from point B into the profile view by constructing a line perpendicular to the F/P fold line. This is shown in Figure 7-13.

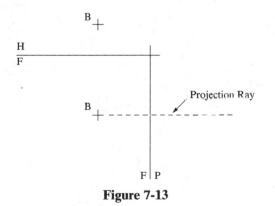

Figure 7-13

We know that point B is located somewhere along this projection ray, according to the rules of orthographic projection. To determine B's location along this projection ray within the profile view, you need to determine the distance from the point to the front plane. Recall that the front plane is seen as an edge in the profile view (as the F/P fold line) and that the front plane is also seen as an edge in the horizontal view (as the F/H fold line). You can use your dividers to measure the distance, D, from B to the frontal plane in the horizontal view and transfer that distance to the profile view. This is illustrated in Figure 7-14, which shows the three orthographic views of point B.

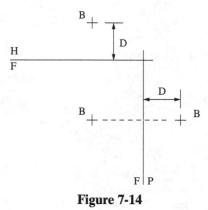

Figure 7-14

Lines in Space. Lines are one-dimensional geometric entities in space. Two points are necesary to define a line, and lines are theoretically infinite in length. In this text, we will represent lines as finite segments defined by their two endpoints.

Using the principals of orthographic projection, you can draw lines in space by connecting their endpoints in each of the principal views. Figure 7-15 shows the horizontal, frontal, and profile views of a line A–B. Notice that the rules of orthographic projection apply. In other words, the points project from one view to another along perpendicular projection rays. The distance between point A and the front plane (labeled "X" in the figure) is seen in both the horizontal and profile views. The distance between B and the profile plane (Y) is visible in the top and front views, and so on.

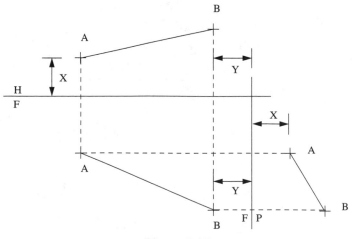

Figure 7-15

When a line is parallel to a plane, say the front plane, it will be seen in its true length in that view. To prove this to yourself, hold a pencil in a plane parallel to the plane of your eyes, so that you are looking at it in its true length. Rotate the pencil away from you while keeping your eyes in the same relative position. Notice that as you rotate the pencil away from you, it begins to seem "shorter," i.e., you are seeing a foreshortened view of the pencil. Now rotate the pencil back so that you are viewing it true length. Rotate the pencil in the plane which is parallel to your viewing plane (i.e., rotate it clockwise or counterclockwise relative to your point of view). If you rotate the pencil in this parallel plane (so that it always remains in the plane), note that it always appears in true length. Thus, it does not matter what the orientation of the pencil is within the plane as long as it lies **in** this parallel plane in order to see the pencil in its true length.

Returning to the discussion of Descriptive Geometry, how can you tell if a line is seen in its true length (TL) within any given view? You know that a line is seen TL if it is parallel to a particular view. Whether or not a line is parallel to a given view is not apparent in the view itself. However, if you look in an adjacent view, it will be parallel to the fold line in the adjacent view. Figure 7-16 shows two lines, AB and CD. Notice that AB is parallel to the H/F fold line in the Horizontal view and parallel to the F/P fold line in the Profile view. This indicates that the line AB is parallel to the front plane and therefore is seen TL in the front view. Line CD is not parallel to any of the fold lines in any of the principal views and therefore is not seen TL in any of these views. Notice how the apparent length of CD changes from one view to the next.

Classification of Lines.   Lines are classified according to whether or not they are parallel to a given plane in the orthographic projection system. For example, a Horizontal Line is parallel to the horizontal plane (and is seen TL in the Horizontal view), a Frontal Line is parallel to the front plane, and a Profile Line is parallel to the profile plane. Thus, line AB shown in Figure 7-16 is classified as a Frontal line. A line which is not parallel to any of the principal views (such as line CD in Figure 7-16) is called an oblique line.

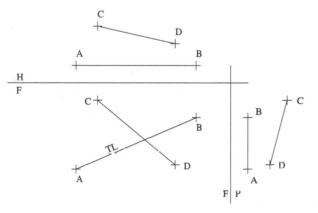

Figure 7-16

It is possible for a line to be parallel to more than one principal view. Figure 7-17 shows three views of a line EF. Notice that in the top view the line is parallel to the H/F fold line, and that in the front view the line appears parallel to the H/F fold line. This indicates that the line is parallel to both the Horizontal Plane and the Frontal Plane. Notice also that the line EF projects into the Profile view as a point. Thus, in the profile view, you are looking down the end of the line. This is known as the Point View (PV) of a line. The classification of a line such as EF, which is parallel to two of the principal planes, is indicated by each of the views to which it is parallel. Thus, EF is classified as a Horizontal–Frontal Line (or a Frontal–Horizontal Line).

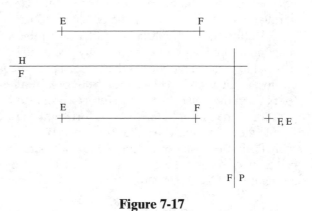

Figure 7-17

True Length and Point View of an Oblique Line.   In many descriptive geometry problems, it is first necessary to obtain a true length (TL) view of an oblique line. This is accomplished by constructing an auxiliary view to which the line is parallel. This is similar to the procedure outlined in the previous sections of this chapter where you constructed an auxiliary view parallel to the edge view of an inclined surface. If you project the line into this parallel auxiliary view, it will then appear in true length (TL). As you project the line into this new view, remember to apply the principals of orthographic projection. The procedure you use to obtain the TL view of a line is outlined in the following and is illustrated in Figure 7-18.

- Draw a fold line which is parallel to either view of the line [labeled H/1 in Figure 7-18(a)].
- Project the endpoints of the line into the auxiliary view along perpendicular projectors [Figure 7-18(b)].
- Measure the distance that these points are from the fold line by looking in the view which is "two views back" [Figure 7-18(c)].
- Connect the endpoints of the line to obtain its TL view [Figure 7-18(d)].

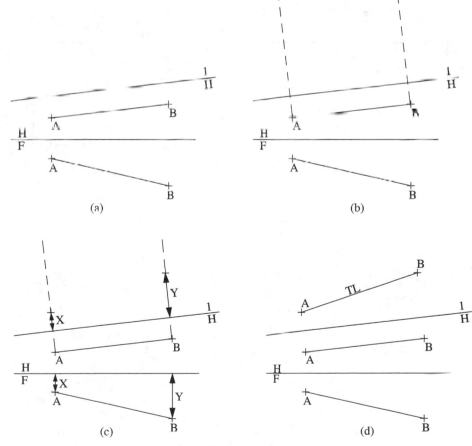

Figure 7-18

The point view (PV) of a line is another useful descriptive geometry construction. The point view of a line is obtained when you are looking down the end of the line. This appears in a view which is perpendicular to a TL view of the line. To prove this to yourself, hold your pencil in a horizontal position so that you are viewing it TL. Rotate the pencil 90° about a vertical axis and you should notice that you are now looking down the end of the pencil—you see the point view of the pencil. An auxiliary view which shows the point view of a line is easily constructed from any view which shows the line in TL. The procedure you follow to obtain the PV of a line is outlined in the following and is illustrated in Figure 7-19.

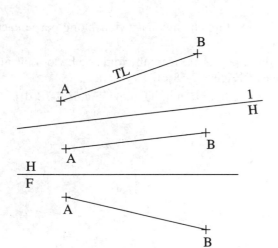

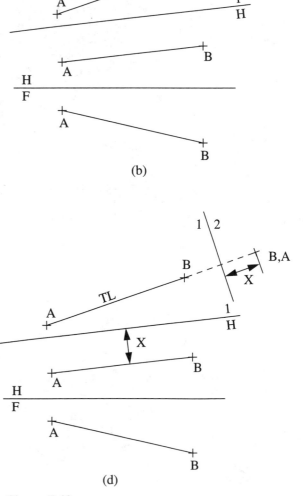

Figure 7-19

- Construct a TL view of the line if necessary [Figure 7-19(a)].
- Draw a fold line perpendicular to the TL view of the line [labeled 1/2 in Figure 7-19(b)].
- Project the endpoints of the line into this new view along perpendicular projection rays [Figure 7-19(c)].
- Measure the distance that the endpoints are from this new fold line by looking in the view which is "two views back." The two endpoints should project as a single point in this view, as illustrated in Figure 7-19(d).

Planes in Space.   Planes are two-dimensional geometric entities which are theoretically infinite in width and depth but which have zero thickness. In descriptive geometry applications, planes are defined by one of four methods:

1)   Parallel Lines.
2)   Intersecting Lines.
3)   A Line and a Point which is not on the line.
4)   Three Non-colinear Points.

Parallel lines will be parallel in all views of the lines. If you can construct a view of the lines in which they are not parallel, then the lines are not truly parallel lines. Figure 7-20(a) shows the horizontal, front, and profile views of lines AB and CD. Notice that in the horizontal view the lines appear parallel, whereas in the front and profile views they appear to intersect. Thus, the lines are not parallel lines. Figure 7-20(b) shows three views of the lines EF and GH. Notice that these lines appear parallel in all three views. If you were to construct an auxiliary view of these lines, they would also appear parallel in that view. Thus, EF and GH are parallel lines and they define a plane in space.

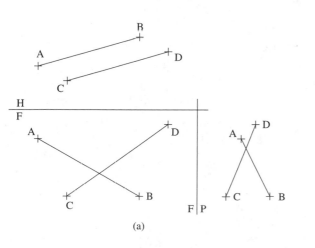

(a)

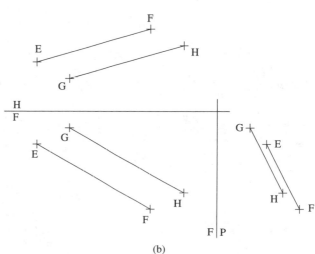

(b)

Figure 7-20

If two lines are intersecting lines, their point of intersection will project from one view to the next. In other words, the point of intersection is a point on each line and, thus, should be on each line in all views of the lines. Figure 7-21(a) shows two lines, AB and CD. These lines appear to intersect in the horizontal view at point X. In projecting X into the front view, you can see that it does not project to the same point on each line. Similarly, the lines appear to intersect at point Y in the front view. If Y is projected into the horizontal view, you can see that it does not project to a single point on both lines. Thus, AB and CD are non-intersecting lines and do not form a plane. AB and CD are known as skew lines (non-intersecting, non-parallel lines). Figure 7-21(b) shows lines EF and GH, which appear to intersect at a point X in the horizontal view. Notice that this point does project to a common point on the two lines in the front view. Thus, EF and GH are intersecting lines and they form a plane in space.

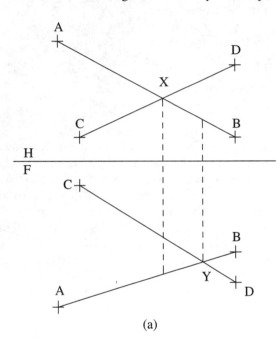

(a)

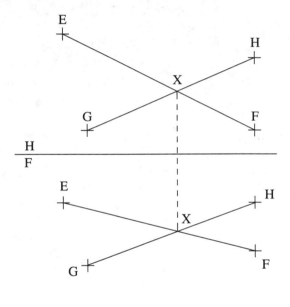

(b)

Figure 7-21

Figure 7-22(a) shows a plane defined by a line (AB) and a point (C) not on the line, and Figure 7-22(b) shows a plane defined by three points (DEF). Notice that, if you connected the point C to the endpoints of the line AB, you would have the exact same definition of a plane as shown in Figure 7-22(b). Notice also that the plane DEF could be thought of as being defined by three intersecting lines—DE and EF intersect at point E (the point of intersection, E, projects orthographically); EF and FD intersect at point F; and DE and FD intersect at point D.

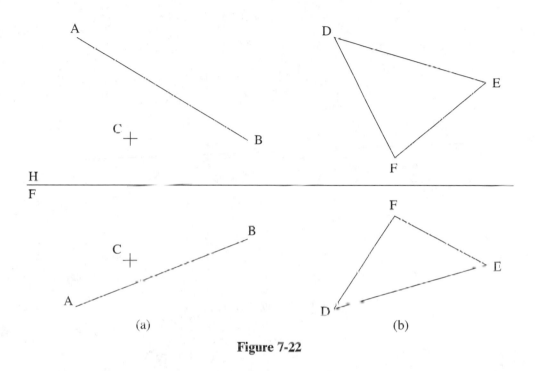

Figure 7-22

When a plane is parallel to a viewing plane, it will appear in true size (TS) in that view. This is consistent with the discussion of normal surfaces from Chapter 5. Recall that normal surfaces were defined as being parallel to one of the principal viewing planes and were thus seen in true size and shape in those views. When a plane is parallel to a viewing plane, say the front plane, it will be seen in its true size in that view. To prove this to yourself, hold a piece of paper in front of you with the paper parallel to the plane of your eyes, so that you are looking at it in its true size. Rotate the paper away from you while keeping your eyes in the same relative position. Notice that as you rotate the paper away from you, it begins to seem "smaller," i.e., you are seeing a foreshortened view of the paper. Now rotate the paper back so that you are viewing it true size. Rotate the paper in the plane which is parallel to your viewing plane (i.e., rotate it clockwise or counterclockwise relative to your point of view). If you rotate the paper in this parallel plane, note that it always appears in true size.

How can you tell if a plane is seen in its true size (TS) within any given view? You know that a plane is seen TS if it is parallel to a particular view. Whether or not a plane is parallel to a given view is not apparent in the view itself. However, if you look in an adjacent view, it will appear as an edge which is parallel to the fold line in the adjacent view.

Figure 7-23 shows two planes, ABC and DEF. Notice that ABC appears as an edge in the horizontal view and that the edge view of the plane is parallel to the H/F fold line in the Horizontal view and parallel to the F/P fold line in the Profile view. This indicates that the plane ABC is parallel to the front plane and therefore is seen TS in the front view. Plane DEF is seen as an edge in the front view, but this edge is not parallel to any of the fold lines. Therefore, it is not seen TS in any view. Notice how the apparent size of DEF changes from one view to the next.

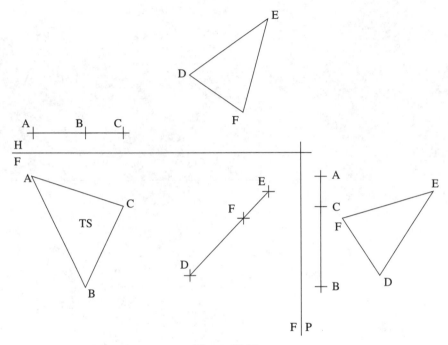

Figure 7-23

In a true size view of a plane, all of the lines on the plane are seen in their true length. For plane ABC in Figure 7-23, all lines on the plane are seen in the edge view as shown in the horizontal view. Since the edge view of the plane is parallel to the H/F fold line, all lines on the plane are also parallel to the H/F fold line in that view. Thus, they are all seen in TL in the front view (the TS view of the plane).

Classification of Planes. Planes are classified according to whether or not they are parallel to a given plane in the orthographic projection system. For example, a Horizontal Plane is parallel to the horizontal viewing plane (and is seen TS in the Horizontal view), a Frontal Plane is parallel to the front viewing plane, and a Profile Plane is parallel to the profile viewing plane. These were referred to as normal surfaces in Chapter 5.

Inclined planes are seen as an edge in one of the principal viewing planes, but the edge view of the plane is not parallel to any fold line. Thus, it is not seen in TS in any of the principal views. Plane DEF in Figure 7-23 is an inclined plane because it is seen as an edge in the front view. This corresponds to the definition of inclined surfaces which you were introduced to in Chapter 5. Oblique planes are not seen as edges in any of the principal viewing planes (recall the discussion of oblique surfaces from Chapter 5).

Edge View and True Size of an Oblique Plane.   Section 7.1 described the procedure you use to obtain the TS view of an inclined surface–you create an auxiliary view which is parallel to the inclined surface, project the surface into that view, and it appears there in TS. The same procedure is followed for an inclined surface in descriptive geometry applications. However, how do you go about creating a true size view of an oblique plane? The first step is to create an edge view of the plane, and from there create a view parallel to the edge view. A plane will be seen edge view when any line on the plane is seen point view. The procedure you use to create an edge view of an oblique surface is outlined in the following and is illustrated in Figure 7-24.

- Determine the location of a TL line on the plane. A TL line is determined by drawing a line from one of the points defining the plane parallel to a fold line. This line will intersect one of the edges of the plane and this point of intersection is projected into the adjacent view (since the lines are in the plane, the point of intersection projects orthographically). Thus, the adjacent view of the line will be true length [see Figure 7-24(a)].

- Draw a fold line perpendicular to the TL view of the line [labeled H/1 in Figure 7-24(b)].

- Project the points defining the plane into this new view along perpendicular projection rays [Figure 7-24(c)].

- Measure the distance that the endpoints are from this new fold line by looking in the view which is "two views back." The points defining the plane should project as a single line in this view, as illustrated in Figure 7-24(d).

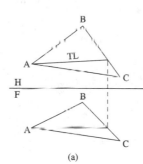
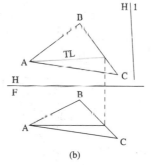

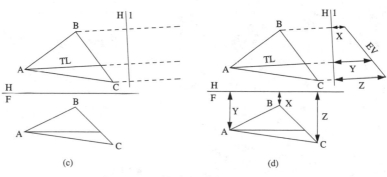

Figure 7-24

Once you have an edge view of the plane, the true size view of an oblique plane is determined by the same method you used to draw the true size view of an inclined plane. This method is outlined in the following and is illustrated in Figure 7-25.

- Draw a fold line which is parallel to the edge view of the plane [labeled 1/2 in Figure 7-25(a)].
- Project the points defining the plane into the auxiliary view along perpendicular projectors [Figure 7-25(b)].
- Measure the distance that these points are from the fold line by looking in the view which is "two views back" [Figure 7-25(c)].
- Connect the points with lines to obtain the TS view of the plane [Figure 7-25(d)].

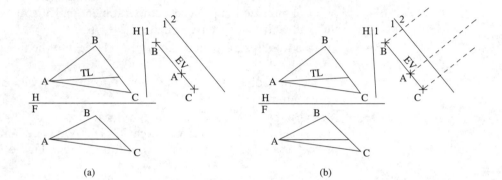

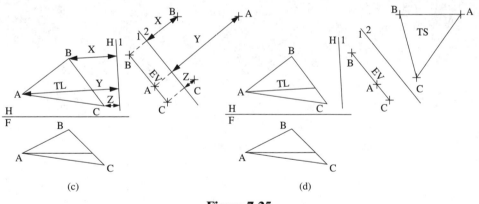

Figure 7-25

Using the principals outlined in this section on descriptive geometry, you can learn to solve various spatial problems. For example, you can determine the point of intersection between a line and a plane or the line of intersection between two planes. Once you know the line of intersection between two planes, you can obtain a point view of that line to determine the angle between two planes (dihedral angle). You can solve for the shortest distance between a point and a line, between a point and a plane, or between two skew lines. Descriptive geometry methods have been used for many years to solve complex spatial problems. Many of these problems can now be easily solved with 3-D models on the computer screen. In fact, the descriptive geometry principles contain much of the foundation or theory behind computer graphics software. Descriptive geometry provides a graphical means for solving geometric problems. The various fields of engineering are full of design problems in which geometric relationships play a huge role. What follows are some sample solved problems, along with additional problems for you to consider.

Car Suspension A-Arms.   Figure 7-26 shows the front and top views of the front portion of an open-wheeled racing car. The suspension members, which run from the wheels to the frame of the vehicle, are referred to as "a-arms" because of their geometry. The design of the shape of the a-arms and their orientation with respect to the vehicle and the ground is a critical aspect of suspension design. Based upon the geometry shown, you are to determine the true length of the tubular a-arm member labeled AB.

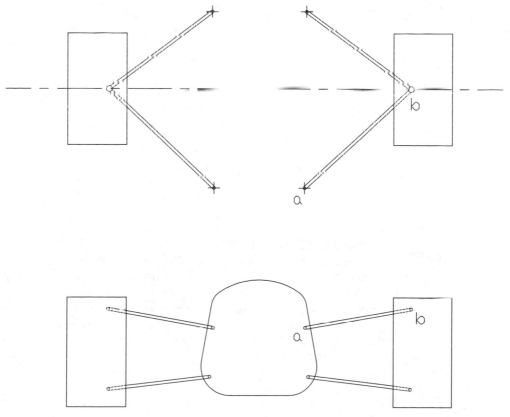

Figure 7-26

This is a classical example of a "true length" descriptive geometry problem. First, you redraw the members in the Front and Horizontal views at the orientations determined by the design. Next, you create a fold line parallel to one of the members. Project the endpoints A and B perpendicular to this fold line and transfer the distances from the other given view (two views back). This process is shown in Figure 7-27. For our example, we chose to project from the Horizontal view of the member and transfer distances from the Frontal view.

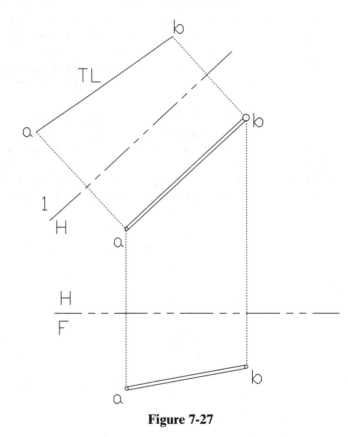

Figure 7-27

Race Car Suspension Clearances. Figure 7-28 shows the front and top views of the design for a vehicle suspension, including the vehicle a-arms and the shock absorber. Note that because of the suspension geometry, the shock absorber must pass through the opening of the upper a-arm. The shock absorber will be purchased from a supplier so it will have fixed size dimensions. The a-arm geometry has been determined by the ride and handling requirements of the vehicle. It is desired to determine the maximum diameter which the tubular a-arm members may have and still allow clearance for the shock absorber.

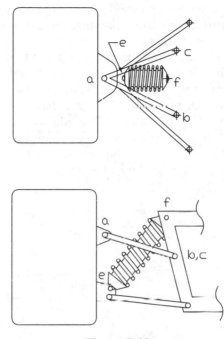

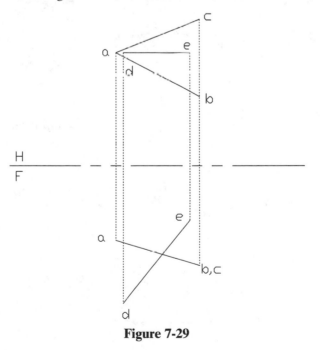

Figure 7-28

To solve this problem, we will determine the shortest distance between the axis of the shock absorber and each a-arm member. The shortest of the two distances is our critical value and may be used to determine the a-arm tube diameter. First, the axis of the shock absorber and the a-arm members are redrawn in the Frontal and Horizontal views. This is shown in Figure 7-29. To determine the shortest distance between two

Figure 7-29

lines, we need to create a view where one of the lines appears as a point. The perpendicular distance from the line to the point view represents the shortest distance between the two lines.

We know from previous material that the point view of a line may be created by projecting in the direction of a true length. Therefore, the usual first step in solving a problem such as this would be to create a view showing one of the lines as true length. However, in this case we already have a true length view of the shock absorber axis (DE) in the Frontal view (how do we know this is true length?). Our next step is to project all three lines in the direction of this true length line. The resulting view will allow us to measure the distance between the a-arm axes and that of the shock absorber. This is shown in Figure 7-30.

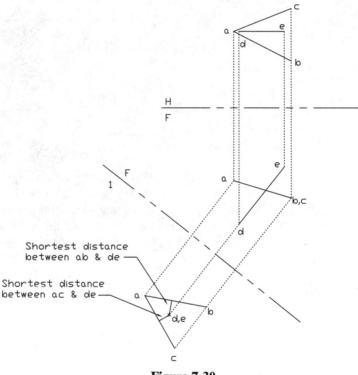

Figure 7-30

EXERCISES 7.3

1. Copy the figure onto grid paper and then construct the third view for the points A–F shown in two views in the figure below.

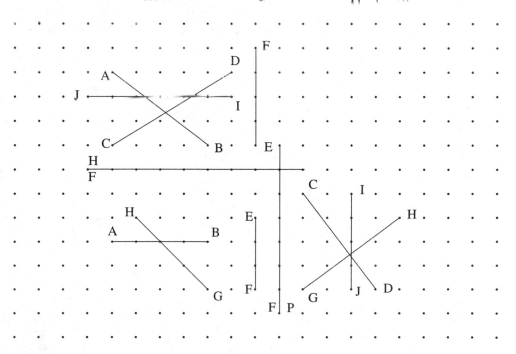

2. Copy the figure onto grid paper and then construct the third view for the lines AB, CD, EF, GH, and IJ shown in two views in the figure below. Classify each line as either Frontal, Horizontal, Profile, or Oblique and label True Length views where appropriate.

3. Copy the figure onto grid paper and then construct true length and point view for the lines.

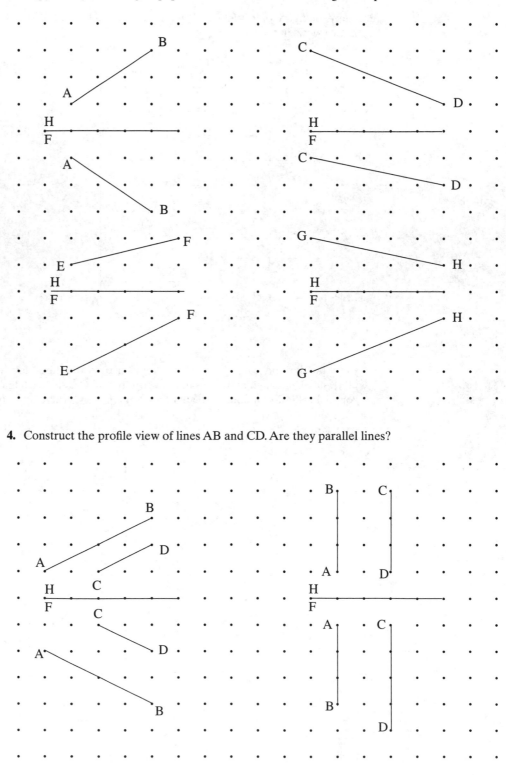

4. Construct the profile view of lines AB and CD. Are they parallel lines?

5. Construct the profile views of lines AB and CD. Are they intersecting lines?

6. Copy the figure onto grid paper and then construct edge and true size views for the planes

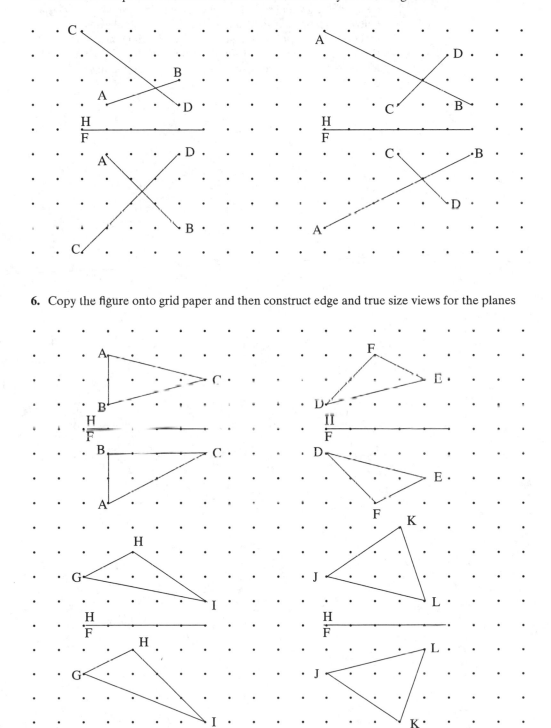

7. A 10′ by 8′ barrier is supported by two diagonal braces which attach to the barrier at a point 6′ above the ground and are anchored at a distance 8′ from the barrier as shown in the figure below. Each brace is to also have a stiffening member which attaches to the support in a tee junction and extends to the ground. The stiffening member is to be located at a point 40% of the length of the support from its ground anchor. How long must the stiffening member be? Recreate the given geometry in a Front and Horizontal view on dot paper using a scale of one dot = 1′.

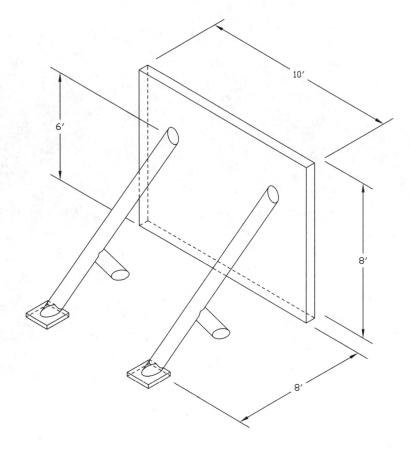

8. The figure shown below depicts a cast housing to which two pipes will attach. One pipe will carry water and the other will carry oil. The locations and lengths of the axes for the inlet and outlet passages are given on the figure. If the oil flow passage is specified to have a diameter of 1.00 inch, what is the maximum diameter the water passage can have and still maintain a 0.125 thickness of material between the passages? Transfer the critical dimensions for the axes to dot paper using a scale of one dot = 1″.

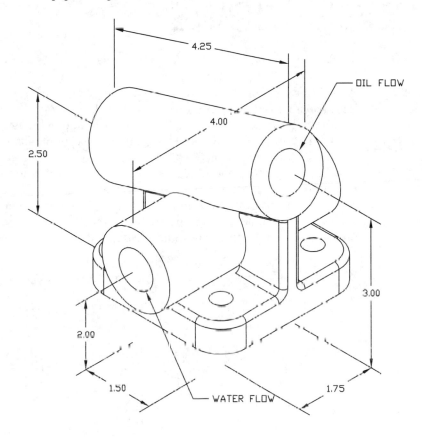

Section Views

• •

As you have no doubt discovered by this time, it is much easier to interpret the visible features in an orthographic drawing than the hidden ones. Section views are included in a orthographic drawing to reduce the number of hidden features in a view. Any of the principal views may be represented "in section" or a supplemental view showing the section may be included. The concept of sectioning is quite simple–it is as if a cutting plane were passed through the solid object and a portion removed to display the object's interior. Figure 8-1 demonstrates the concept of a section view. The material through which the cutting plane passes is shown as crosshatched. An easy analogy is to imagine that the component were actually cut using a saw, and the marks left by the motion of the saw blade in cutting the material would be simulated by the crosshatching.

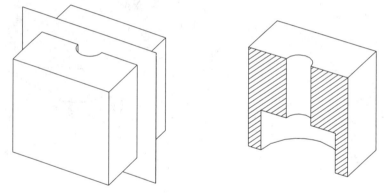

Figure 8-1

Crosshatching is a pattern of lines (or symbols in some civil engineering applications) constructed at a nominal angle between 30 and 60 degrees from the horizontal and with uniform spacing. The default format for crosshatching in most mechanical and civil applications is a pattern of continuous diagonal lines. The lines should be evenly spaced with an interval distance of approximately 3/32 (2.5 mm) and typically at an angle of 45 degrees. Figure 8-2 shows an example of the angle and interval spacing of typical cross-

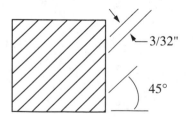

3/32"

45°

Figure 8-2

hatching. A different angle may be used if too many of the view's object lines are also at 45 degrees or for some specific mechanical engineering applications (see Section 8.3). In both civil and mechanical engineering applications, different hatch patterns may be used to represent different materials. Examples of the various standard hatch patterns are shown in Figure 8-3.

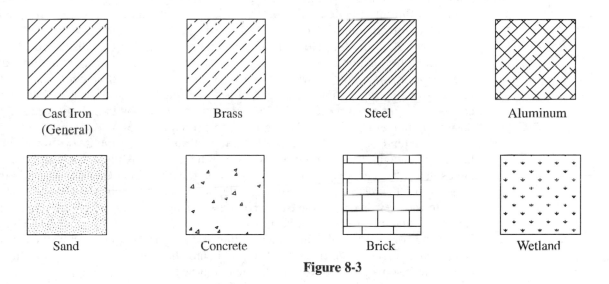

Cast Iron (General)   Brass   Steel   Aluminum

Sand   Concrete   Brick   Wetland

Figure 8-3

There are some guidelines you should follow when adding crosshatching to a drawing either by computer or by sketching. 1) Crosshatch lines should never be parallel to any of the lines bounding the area to be crosshatched. 2) Crosshatch lines should be thinner than visible lines so they will not be confused with the object lines. 3) Crosshatch lines for separate portions of a single object are angled in the same direction. Crosshatch lines which go in different directions denote that there is more than one object being sectioned (as in an assembly). These rules for crosshatching are illustrated in Figure 8-4.

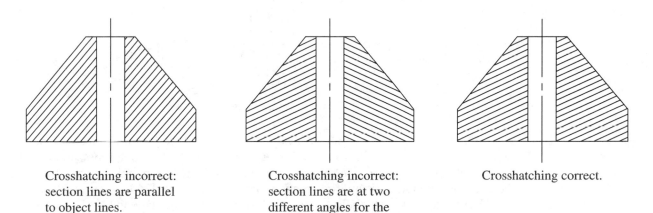

Crosshatching incorrect: section lines are parallel to object lines.

Crosshatching incorrect: section lines are at two different angles for the same part.

Crosshatching correct.

Figure 8-4

No hard and fast rules exist as to when a section view should be included in a drawing. It is up to the individual creating the drawing to make this decision. It is not uncommon for a drawing to include several section views in order to make the drawing easier to read.

Section views are very common in both the mechanical and civil engineering disciplines. The standards of engineering graphics specify different types of sections to be used in different situations. The seven basic types of section views are full, half, broken-out, revolved, removed, offset, and aligned.

Very often, the line view of the imaginary cutting plane is displayed in the view from which the section is projected. This is referred to as the cutting plane line and uses one of two formats, both of which differ from the conventional object line. The two preferred formats are: 1) a line consisting of a set of equal length dashes, or 2) a line which consists of a repeated pattern of a long segment followed by two dashes. Examples and desired actual lengths for the dashes are shown in Figure 8-5. As seen in the example, the line will extend beyond the edges of the object and will terminate with perpendicular lines (approximately 1/4 inch in length) which have arrowheads at their end. The arrowheads of a cutting plane line are used to indicate the direction of visibility used in the construction of the section view. For example, arrowheads which point from right to left indicate the view is being constructed as viewed from the right side. Remember that all standard rules of projection still apply in the placement of section views. You cannot ignore the conventions of projection by simply reversing the direction of the arrows. The rules of precedence state that when a cutting plane line and a center coincide, the cutting plane line is shown and not the centerline (as shown in Figure 8-5).

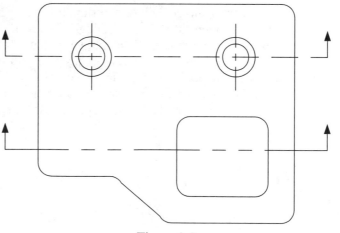

Figure 8-5

A cutting plane line is not always included in the section drawing. As a rule, all removed, offset, and aligned sections should have a cutting plane line represented. In the case of full and half sections, a cutting plane line need only be included if the location of the cut is not obvious or if multiple sections are being constructed in the same drawing.

Very often, a set of labels are included with the cutting plane line and with the sectional view. These labels consist of a pair of uppercase letters at the endpoints of the cutting plane line and a name using the same letters below the crosshatched view. The same letter is used for both endpoints of the same cutting plane line, and the section view name includes both letters. For example, if the endpoints are both labeled 'A,' the sectional view label would be 'SECTION A-A'. As shown in Figure 8-6, the sectional view name includes the word "section" and the two letters corresponding to the endpoints separated by a dash.

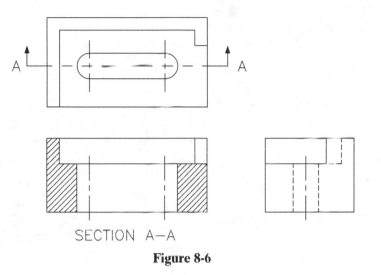

SECTION A—A

Figure 8-6

8.1 *SECTION TYPES*

Full Section.   Probably the most common type of section is the full section. In a full section, the cutting plane passes entirely through the object. The illustration in Figure 8-1 is an example of the concept of a full section. Figure 8-7 shows the same object in an orthographic projection, with the front view in full section. Note that hidden lines are usually omitted in the sectioned view, although they may be included on occasion for purposes of clarity or dimensioning. You should also notice that all visible features beyond the cutting plane are included in the section view. This is typical of most section types.

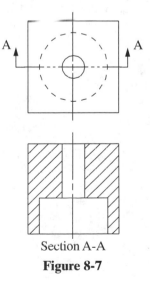

Section A-A

Figure 8-7

Half Section. In a half section the theoretical cutting plane passes half-way through the object, with a perpendicular cut made to remove one quarter of the material of the object. Figure 8-8 shows an example of this. In general, half sections will show

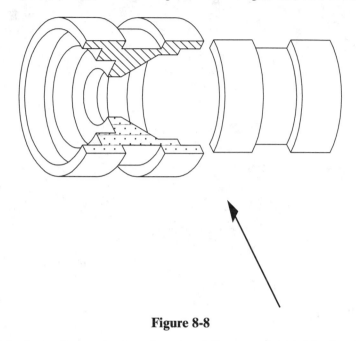

Figure 8-8

both internal features (in section) and external features (as visible lines) in the same view. Half sections are typically used for symmetrical components since a single view will display both visible and hidden features, thus reducing the number of views which must be constructed. In the case of a half section constructed from a symmetrical object, you need not include hidden lines in the non-hatched half of the view since these features are depicted in the cross-hatched portion, as shown in Figure 8-9. The hidden lines may sometimes be included in the interest of clarity or for purposes of dimensioning. The cutting plane line for the half section only requires one arrowhead on the portion of the line perpendicular to the line of projection. The crosshatching in the sectioned view terminates at the centerline in the view or, if none is present, an object line may be added.

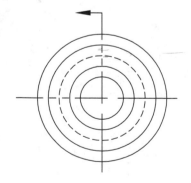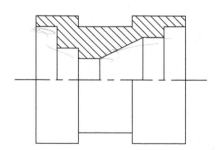

Figure 8-9

Broken-out Section.   A broken-out section is often used in place of a full or half section. In cases where the clarity of only a single feature need be enhanced or only a small portion of the object needs to be sectioned, a broken-out section may be used instead of a full or half section. The broken-out section permits you to only section the portion of the drawing where more detailed information is desired. The section area is bounded by a free-hand "break line." A broken-out section is shown in Figure 8-10.

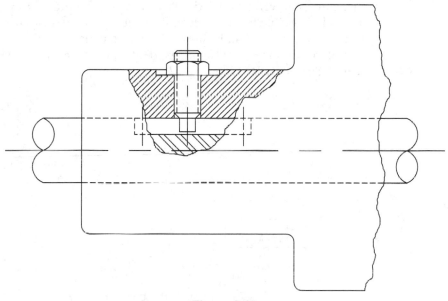

Figure 8-10

Removed Section.   In the case of a removed section, a cutting plane is passed through the object and a planar "slice" is taken. The slice is rotated about an imaginary axis which lies within the plane of the slice. Figure 8-11 shows an example of the orientation of the cutting plane and the axis of rotation. The "slice" is rotated 90 degrees in

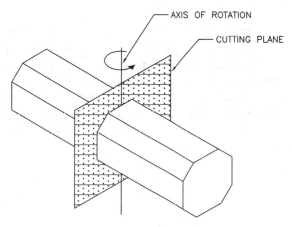

Figure 8-11

order to display the section true size. The sectioned area is then placed outside of the object. This location may be in any open space within the sheet of paper or even on another sheet. The latter is very common in architectural graphics. Because of this, a removed section should always be labeled. In mechanical engineering, removed sections are often used to provide an enhanced description of a cross section not well represented in a conventional view. This is often the case with long objects which change in cross section along their length. Figure 8-12 shows two examples of removed sections. In most cases, the removed section will only show the material cut by the cutting plane and will not include visible features beyond the cutting plane as was the case with the full and half sections. Section A-A in Figure 8-12 is an example. For mechanical drawings, the orientation of the sectioned area should be perpendicular to the line view of the cutting plane as seen in Section B-B of Figure 8-12.

SECTION A–A   SECTION B–B

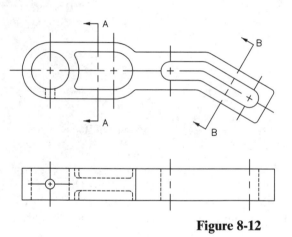

Figure 8-12

Revolved Section.   The revolved section is very similar in appearance and application to the removed section. The primary difference is that the section is placed directly on the source view. The revolved section does not employ a standard cutting plane line but rather uses a centerline to indicate where the cut was taken. The object lines adjacent to the section may be broken to enhance the view as shown in Figure 8-13 but this is not required. In the case of a revolved section taken from a tapering geometry, the section represents the size of the object at the point of the section only.

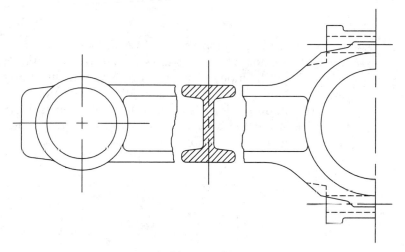

Figure 8-13

Offset Section.   The offset section is a variation on the full section. It is used in cases where the features to be sectioned do not lie along a single line. Figure 8-14 graphically depicts the concept of an offset section. The offset section allows the cutting plane

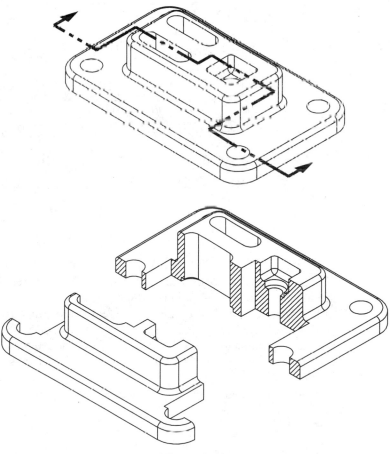

Figure 8-14

to be divided into multiple segments, which are then offset in a direction perpendicular to the cutting plane. This allows the segments of the cutting plane to pass through features which do not lie along a single straight line. The offset portions of the cutting plane line are connected with right angle lines. When the section is constructed, the offset segments are brought into a single plane and then projected as if a full section. In this way, no visible edges from the offsets are shown in the crosshatched view. Figure 8-15 shows an example of an offset section.

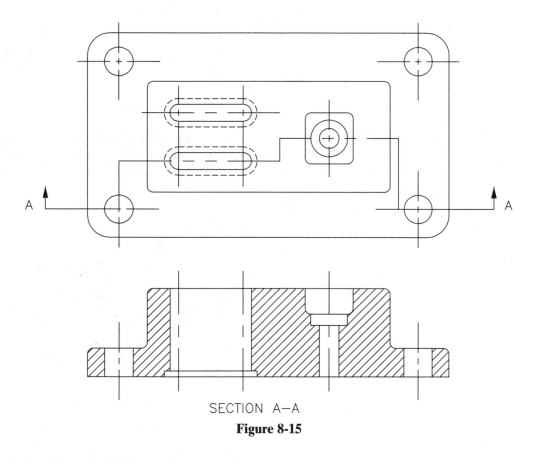

SECTION A—A

Figure 8-15

Aligned Section.   Many objects include features such as holes and spokes, which are arranged in a radial pattern. The orientation of these features may be such that a full or half section would either miss these features or show them as distorted due to the projection direction. The aligned section technique allows the cutting plane to be divided and the segments to be displaced radially in order to pass through these features. As with the offset section, the concept is then that the displaced segments of the cutting plane are aligned and then projected as if it were a full section. Figure 8-16 graphically depicts the concept of an aligned section. In this fashion, the affected features will be shown both in

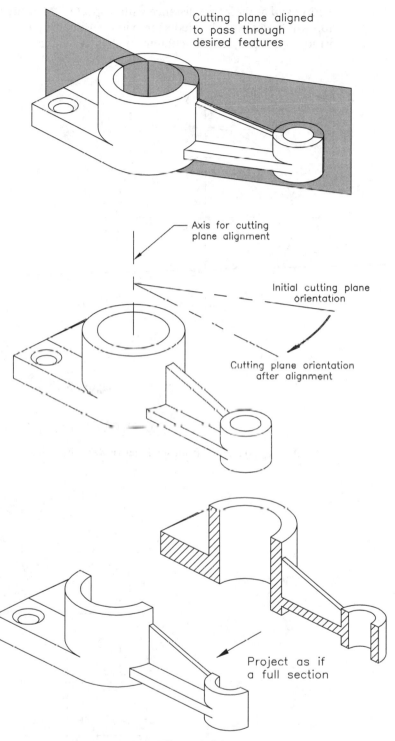

Cutting plane aligned
to pass through
desired features

Axis for cutting
plane alignment

Initial cutting plane
orientation

Cutting plane orientation
after alignment

Project as if
a full section

Figure 8-16

section and at their true distance with respect to the center of the object. The aligned section will typically be labeled as previously discussed. Figure 8-17 shows an example of an orthographic projection including an aligned section.

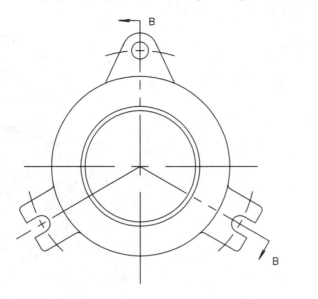
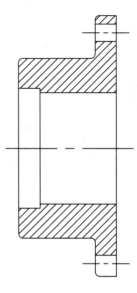

SECTION B—B

Figure 8-17

EXERCISES 8.1

1. Draw a top and full front sectional view for the object shown below.

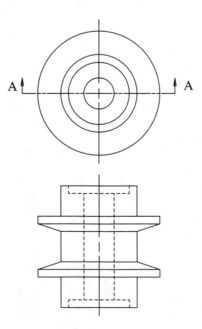

2. Draw a top view and a half-sectional view front view for the object shown below.

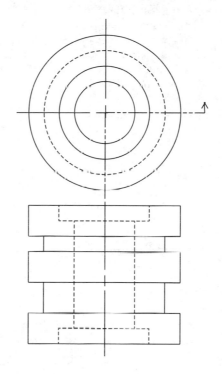

3. Draw a top view and a half-sectional view front view for the object shown below.

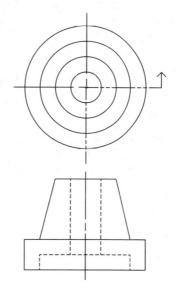

4. On a separate sheet of paper, sketch the top view and a revolved section along line A–A.

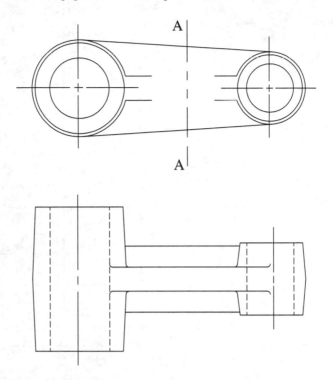

5. Given the top and front views of the object shown below, on a separate sheet of paper draw the removed section A–A, B–B, and C–C. Note: Centerlines have been omitted from this drawing for clarity.

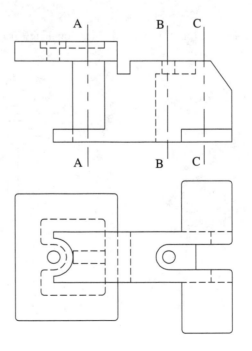

6. In the problems shown below, the views indicated by the balloons are to be changed to full sectional views taken along the center line in the direction indicated by the arrows in the remaining view. For each, select the correct answer from the 24 proposed views shown at the right.

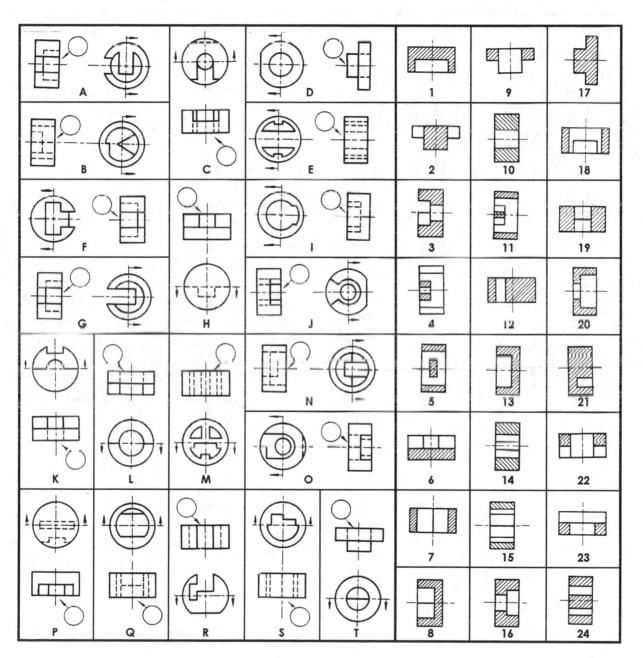

7. In the problems shown below, the views indicated by the balloons are to be changed to half-sectional views as indicated by the letters A–A taken along the center line in the remaining view. For each, select the correct answer from the 24 proposed views shown at the right.

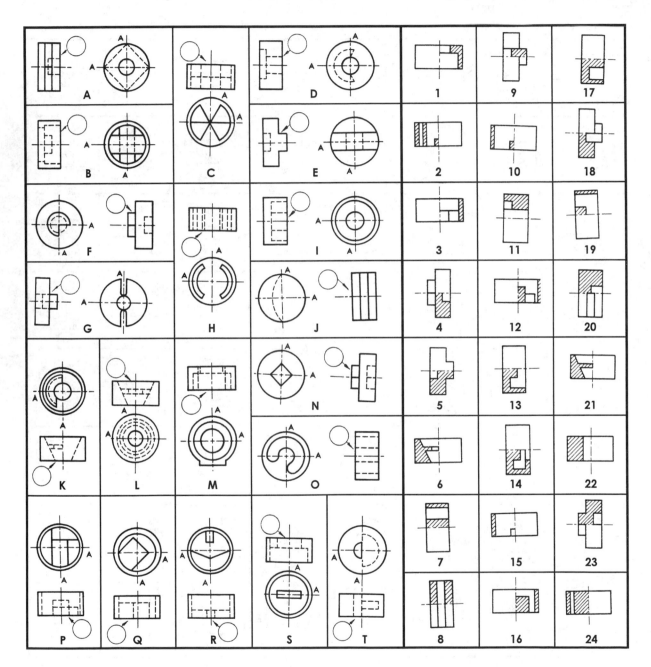

8. In the problems shown below, select the section view to complete each problem from the 30 proposed views shown.

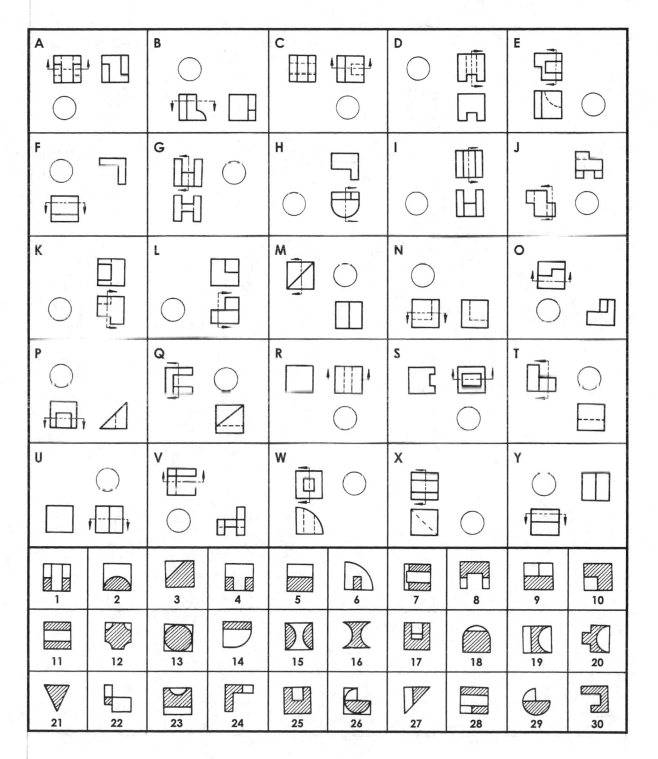

9. In the problems shown below, select the section view to complete each problem from the 30 proposed views shown.

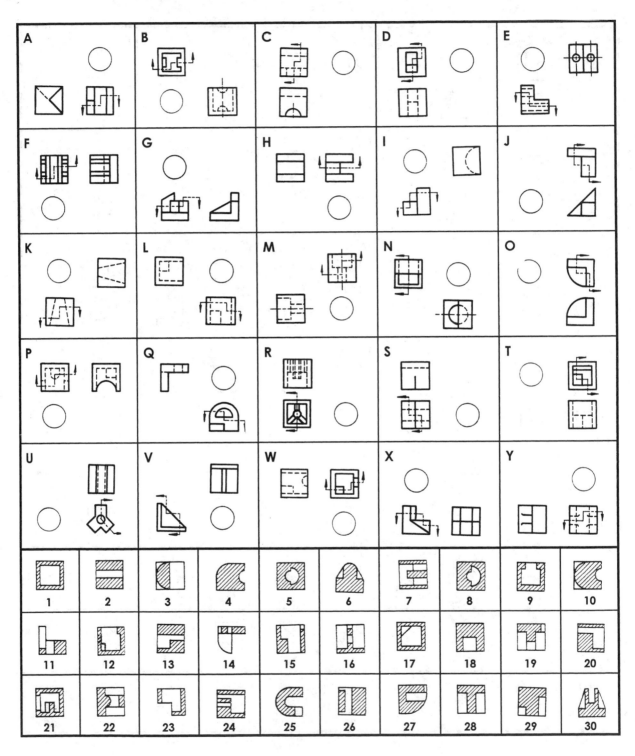

8.2 RIB FEATURES IN SECTION

Many mechanical components, especially those produced by forming processes such as casting, often incorporate features such as ribs and webs to increase strength without adding excess material. Figure 8-18 shows a cylindrical component with a radial pattern of four ribs. Sectional views taken from an object with rib features can present a special

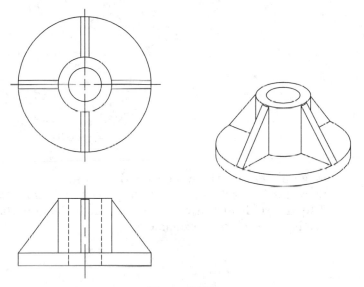

Figure 8-18

problem. If the cutting plane for the section bisects the rib along its length, the standard rules of sectioning should not be followed. To do so would give a misleading representation of the object. This is shown in Figure 8-19.

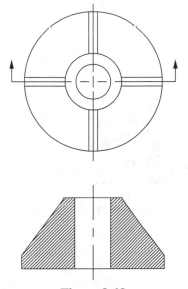

Figure 8-19

The section view shown in Figure 8-19 appears to be that of the object in Figure 8-20 rather than of the object shown in Figure 8-18. In a case such as this, the preferred technique is to not crosshatch the rib features as shown in Figure 8-21. This same standard of not crosshatching rib features when cut along their length is also applied to any

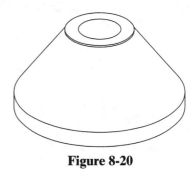

Figure 8-20

long thin feature where showing with crosshatching gives a false impression of material thickness. This rule was illustrated previously in this text in the object shown in Figure 8-16. If a rib feature is cut across its length, then the standard rules of sectioning apply. See Figure 8-22 for an example of this.

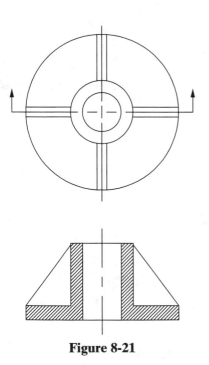

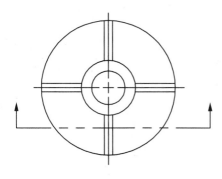

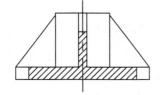

Figure 8-21   **Figure 8-22**

EXERCISES 8.2

1. Given the top and front views of the object shown below. Sketch the top view on a separate sheet and complete the side view (it has been started for you) as a section along A-A. Note: Centerlines have been omitted from this drawing for clarity.

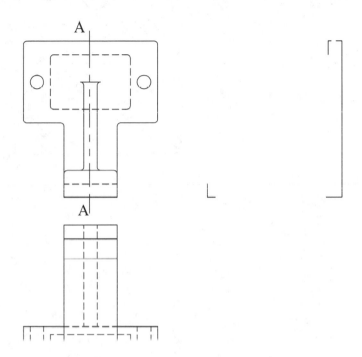

8.3 APPLICATIONS

Mechanical Engineering. A very common drawing in mechanical engineering is the assembly drawing. As opposed to the single part drawings which we have been examining so far, the assembly drawing depicts multiple components. As the name implies, the drawing will show these components in an assembled state. The assembly drawing is often a single profile view of the system, drawn in full section. Special conventions apply when assembly drawings are shown in section. Several of these conventions involve the crosshatching pattern and orientation. In an assembly section, the direction and/or angle of the crosshatching is changed for adjacent parts. The orientation is constant for each individual part, regardless of whether the part appears as separate areas in the drawing. Additionally, the specific sectioning patterns for the material of the part may be used to increase the amount of information given graphically in the drawing. Very thin components such as gaskets and washers would be very difficult to hatch due to their size. In assembly sections, features such as these are not hatched but rather "filled" so as to appear solid. Finally, certain standard parts such as shafts and fasteners are not cross-hatched but rather simply shown as visible material. Figure 8-23 shows a portion of assembly section in which these conventions are represented.

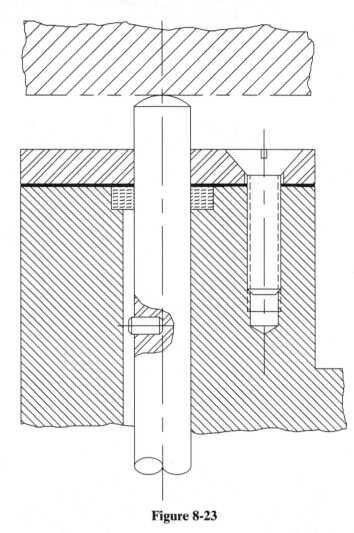

Figure 8-23

Highway Engineering.   When engineers construct paved highways, they need to build a flat driving surface. The ground surface at a highway location generally is not flat; therefore, earth must be removed in some places and built up in others to achieve a flat highway surface. Once the earthwork is complete and a fairly level surface is obtained, pavement can be applied, resulting in a smooth driving surface.

Figure 8-24 shows a cross-section of the ground surface at a given location in a highway construction project. The desired finished road surface is superimposed on the drawing of the ground surface. The finished roadway includes sideslopes for the transition between pavement and existing ground on each side of the road.

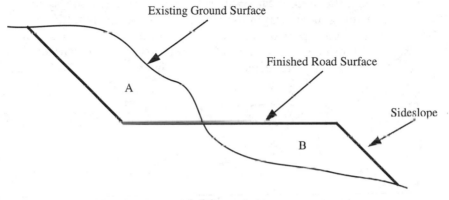

Figure 8-24

As can be seen in this figure, the material in the A region must be removed, or *cut*; whereas, the material in region B must be built up, or *filled*, to achieve the desired road grade.

The engineer needs to know the volume of earthwork for a given construction project so plans can be made for its successful completion. For example, if there is a great deal of excess cut material, the engineer must decide where to "waste" the excess. Conversely, if there is excess fill, a source for the extra material required must be determined. For designing the roadway, the engineer typically has cross-sectional views of the road way at intervals of 50–100 feet. The volume of cut and the volume of fill between two cross-sections is illustrated in Figure 8-25.

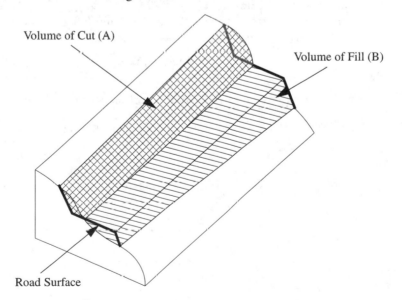

Figure 8-25

In this figure, the volume of cut between the two cross-sections is labeled A, and the volume of fill is labeled B. To compute the volume of cut or fill, you must first determine the approximate area of cut or fill at each cross-section. For example, Figure 8-26 shows a cross-sectional view of the roadway drawn on square grid paper. In this example, to obtain the area of cut, each grid square represents one square foot. You can count all of the full grid squares in the cut area and approximate the area in the partial squares to arrive at an approximate area of cut for this cross-section. Using the same procedure, you can determine the area of fill at this cross-section. In this example, the approximate area of cut is 35 square feet and the approximate area of fill is 25 square feet.

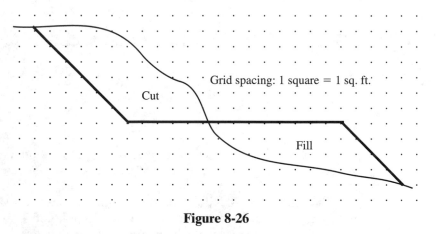

Grid spacing: 1 square = 1 sq. ft.

Cut

Fill

Figure 8-26

To obtain the *volume* of earthwork between two cross-sections, you compute the average area of cut and the average area of fill between the two sections and multiply by the distance between them.

The cross-section shown in Figure 8-27 is located 100 feet away from the previous cross-section (shown in Figure 8-26). The approximate area of cut and fill at this cross-section are 15 square feet and 33 square feet, respectively. Therefore, the volume of cut in this section of the highway is given as:

$$\text{Vol}_{\text{Cut}} = \left(\frac{35+15}{2}\right) \times 100 = 2500\,\text{ft}^3 = \frac{2500}{27} = 93\,\text{yd}^3$$

and the volume of fill in this section is given as:

$$\text{Vol}_{\text{Fill}} = \left(\frac{25+33}{2}\right) \times 100 = 2900\,\text{ft}^3 = \frac{2900}{27} = 107\,\text{yd}^3$$

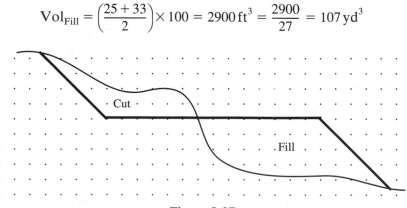

Cut

Fill

Figure 8-27

EXERCISES 8.3

1. The distance between cross section 1 and cross section 2 shown below is 100 feet. Compute the volume of cut and the volume of fill for this section of highway.

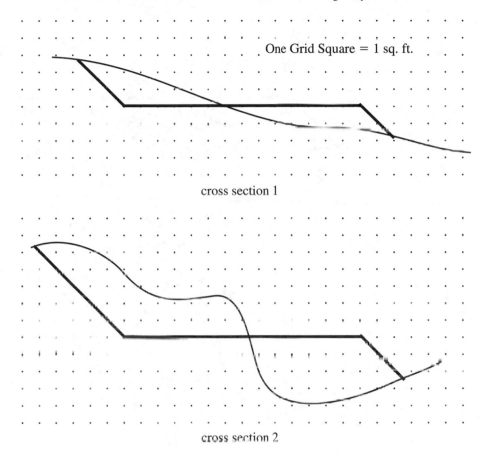

One Grid Square = 1 sq. ft.

cross section 1

cross section 2

2. The distance between cross section 1 and cross section 2 shown below is 50 feet. Compute the volume of cut and the volume of fill for this section of highway.

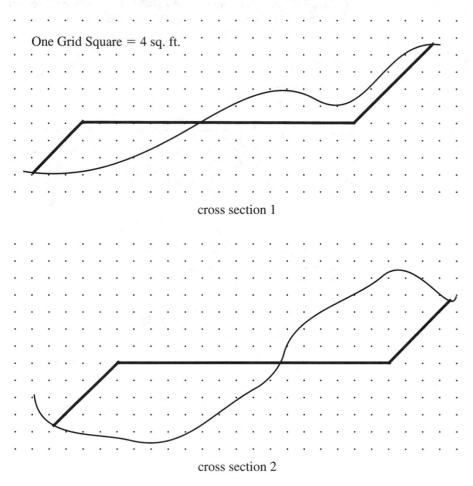

One Grid Square = 4 sq. ft.

cross section 1

cross section 2

C H A P T E R 9

Dimensioning

· ·

An engineering drawing provides a graphical representation of an object. However, for the purposes of construction or manufacture, the graphical representation alone is not sufficient. There are many aspects of an object which cannot be conveyed through just the drawing. The type of material, smoothness of a surface, and special surface treatments are all examples of such information. In addition, the sizes of components, while given graphically in the drawing file, may be open to interpretation or mistake. If a person reading the drawing is required to measure all values from the drawing, they may place a personally biased emphasis on certain features (for example, should a certain value be measured as a diameter or a radius). In the case of most civil and architectural drawings, the drawings are not actual size because that would be extremely impractical. Rather, the drawings are created at some scale where each unit of measurement on the drawing is representative of some real-world distance as discussed in Chapter 6. In this case, measuring distances directly from the graphical representation would be tedious and prone to error.

In order for a drawing to be used for construction or fabrication purposes, some means must be provided for specifying the sizes of the object's features. In engineering graphics, this technique is referred to as dimensioning. Whether the drawing is of a mechanical component from an automobile, the floor plan of a house, or a design for a highway bridge, dimensions must be included in order for the designs to be fabricated. Dimensions specify the sizes and locations of features which comprise an object. They also serve as a set of instructions for the manufacturer about what is contained in the drawing. In addition, the dimensions of an engineering drawing provide a benchmark against which completed objects may be compared to verify their accuracy.

The principles used for the generation of dimensions may be divided into two major categories: the rules which define the format of the dimensions and their placement on the drawing, and the process for determining which dimensions to include for proper fabrication and inspection of an object. The remainder of this chapter will discuss each of these categories in dimensioning practice in more detail.

9.1 FORMAT OF DIMENSIONS

Just as our language has rules of grammar and spelling, the language of dimensioning has rules of format and placement. The form of dimensions must conform to a standard so that anyone who has been trained in engineering graphics can read a drawing created by someone else. These standards are established by a body known as ANSI (American National Standards Institute). The ANSI standards serve as a guide to industry for accepted practice. Most companies adhere to the majority of ANSI standards but may vary from some due to past practices. Some industries are required to use ANSI standards for work performed as part of government contracts.

All letters and numbers included on the drawing must be of a standard form and height to ensure readability. Different types of dimension formats are used for the dimensioning of different types of features. Dimensions may be categorized as linear, angular, cylindrical, and radial. The most common dimension format is the linear dimension.

A linear dimension is used to specify the linear distance between two points on the drawing. This can be the length of an edge or the distance between the centers of two holes. A typical linear dimension will consist of extension lines, dimension lines, and a dimension value. Extension lines (also known as witness lines) are lines which extend from the object features to provide a reference to which the dimension is "attached." The extension lines should be offset from the object points that they reference by a small amount to provide a visible gap and should extend about 1/8 inch beyond the arrowheads of the last dimension line. Both object lines and centerline lines may be used in place of one or both extension lines under certain circumstances. Dimension lines are drawn perpendicular to and touching the extension lines. They typically terminate in two arrowheads pointing in opposite directions and are used to indicate graphically the size range of the dimension and the direction of its application. The value of the dimension is typically placed in a break in the approximate center of the dimension line. Figure 9-1 shows a typical linear dimension with the extension lines, dimension lines, extension line offsets, and dimension value indicated.

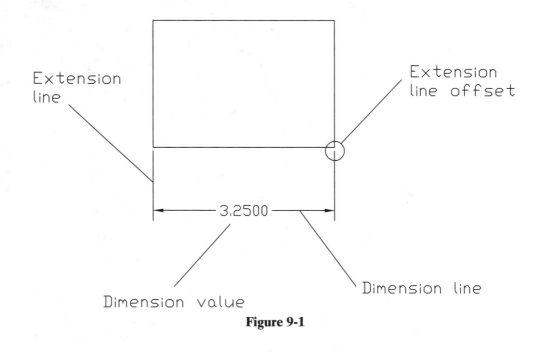

Figure 9-1

In some applications, different types of arrowheads may be used for dimensions as illustrated in Figure 9-2. In particular, slashes are typically used in place of arrowheads when sketching. The slashes are faster and easier to sketch than arrowheads.

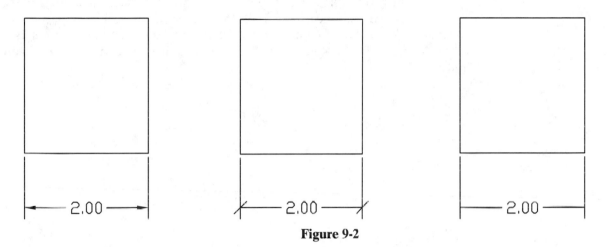

Figure 9-2

The typical format shown in Figure 9-1 is used as long as there is enough space for the arrowheads and dimension values to be drawn at their proper size and for visible dimension lines to be created (normally at least 1/8 beyond the arrowhead). As the distance being dimensioned becomes too small for the default format to be used, alternative forms must be employed. The first alternative employed is to move the dimension lines outside of the extension lines, leaving the dimension value inside. In this case, the arrowheads will point toward each other [Figure 9-3(a)]. As the space becomes too small for the dimension value to be placed inside of the extension lines, the format shown in Figure 9-3(b) may be used.

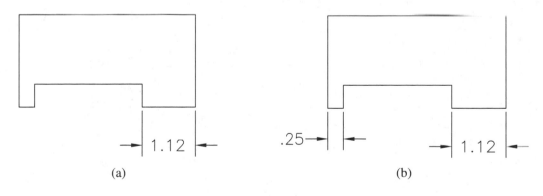

(a)   (b)

Figure 9-3

Angular dimensions are used to specify the angle between two features which meet at some orientation other than perpendicular. As seen in Figure 9-4, the dimension lines used for angular dimensions are arcs with their centers at the vertex of the angle being specified. The text for the angle should be horizontal and placed in a break in the dimension line. As with the linear dimension formats, as the angle being specified decreases, alternative forms should be used to maintain readability. Figure 9-5 shows examples of these alternative forms. Note the similarity between these and the forms of Figures 9-3(a) and 3(b).

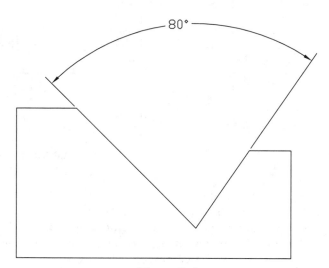

Figure 9-4

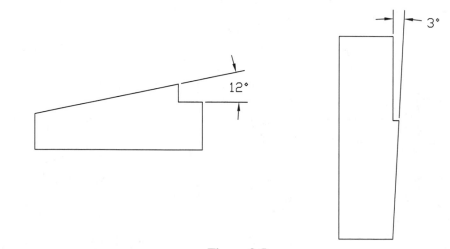

Figure 9-5

Cylindrical form features are non-hole single-curve surfaces. These features should be given as a diameter in cases where the person who would perform the manufacture of the component would make their measurement as a diameter. This is typically the case for single-curve surfaces greater than semi-circular. In cases where a rectangular view projection of the surface exists, it is preferable to place the diametral dimension in this view. This is especially true in the case of multiple concentric cylindrical features. Since this specification is often applied to the rectangular view where the curvature of the surface is not evident, the dimension value is supplemented with a designation to indicate that it is in fact a diametric value. Earlier standards specified a suffix of **D** or **DIA** be added to the value, while more current standards specify the universal diametric symbol (Ø) be used as a prefix to the value. Examples of diametral dimensions are shown in Figure 9-6.

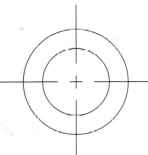
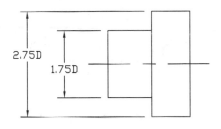

(a)

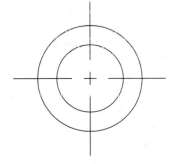

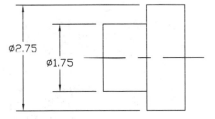

(b)

Figure 9-6

Radial dimensions are used to specify the size of single-curve features which sweep angles of 180 degrees or less. Radial dimensions should always be placed on the circular view of the surface. The radial dimension line should always be drawn at an angle, never horizontal or vertical. The standard form for a radial dimension is that of a dimension line extending from the center of the arc to the arc line and having one arrowhead at its outer end, touching the arc line. The dimension value is placed in a break of the extension line. Figure 9-7(a) shows this format. As the radius size becomes too small for this format, or interference with adjacent dimensions occurs, the modified leader forms in Figure 9-7(b) may be used. Note that all radial dimensions should include the designator R to indicate the specification is a radial value. The center of the radius is usually indicated with a '+' (plus) symbol.

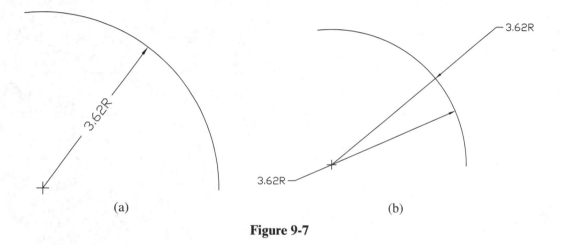

(a)   (b)

Figure 9-7

Possibly the most common occurrence of single curve features in an engineering drawing is in the form of a hole. Holes in mechanical and civil engineering components are used in conjunction with fasteners for aligning and joining different parts. Holes are specified differently in engineering drawings than are cylindrical shape features. Holes are specified in terms of something referred to as a local note or a 'callout'. The local note provides the specifications of the hole and references it by means of a dimensioning tool called a leader line. A leader line is a special type of dimension line, with an arrowhead at one end and a string of text at the other end. The text is the body of the note and refers to the entity at which the arrowhead is pointing. The leader line is drawn at an angle to the horizontal of the paper, preferably at an angle between 30 and 60 degrees. The leader line should never be drawn horizontal or vertical. An arrowhead is placed at the end of the leader line in contact with the specified entity. The other end of the leader line has a horizontal extension of approximately one quarter inch. The leader line will meet the text of the note halfway through the height of either the first character in the note or the last. Figure 9-8 shows examples of leader lines with their respective notes.

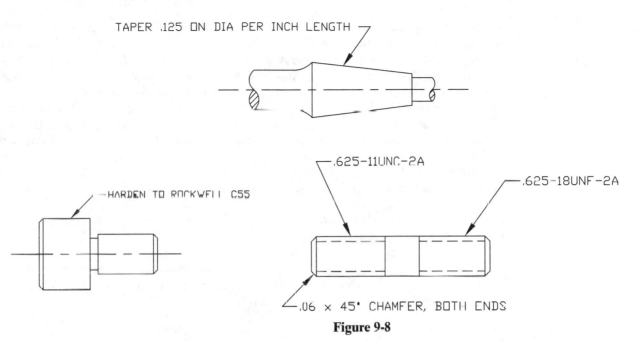

Figure 9-8

When a leader line is specifying the size of a hole, it is preferable to reference the circular view of the hole. The leader line should point at the center of the circle and stop at the outermost diameter if concentric hole features exist. This is shown in the examples of Figure 9-9. These concentric hole features represent the result of a series of manufacturing processes. The leader is pointing at the representation of the last process performed. Figure 9-9 includes several examples of local notes for various machined holes. The hole labeled Example #1 in this figure is a note specifying a hole with internal threads. This hole was produced by first drilling a specific size pilot hole and then cutting

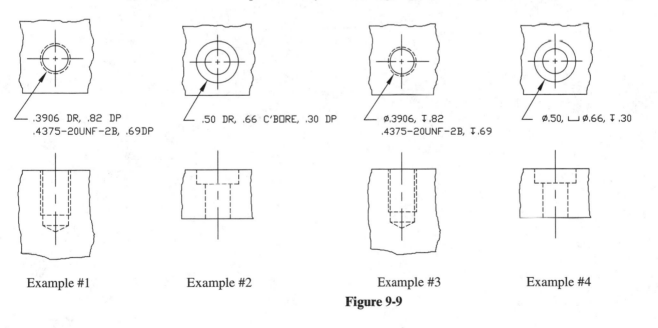

Figure 9-9

threads into the sidewall of the drilled hole. The first line of the note gives the diameter and depth of the drilled hole, and the second line specifies the form of the thread to be cut and the depth to which it is to extend. The second hole in Figure 9-9 (Example #2) shows a note detailing a counterbored hole. A counterbored hole is used to accommodate the head of a mechanical fastener. A counterbored hole consists of a drilled pilot hole and a second bored hole of a larger diameter which is concentric to the drilled hole. The counterbore is used to recess the head of a mechanical fastener below the surface of the material in which the hole is placed. Newer standards for the writing of local notes includes an expanded use of symbolic entities for the various aspects of the note. Examples #3 and #4 of Figure 9-9 show the same notes as #1 and #2 written using the newer symbolic techniques. You may encounter both techniques during your engineering career.

In addition to the local note describing the size of the hole, the location of the hole on the object must be specified. This is usually done with a set of linear dimensions which reference the centerlines of the hole. These dimensions locating the hole should also be placed in the view which shows the hole as circle. This makes sense, since it will take two dimensions to specify the location of the hole and two centerlines are shown for the hole in its circular view. An example of this is shown in Figure 9-10.

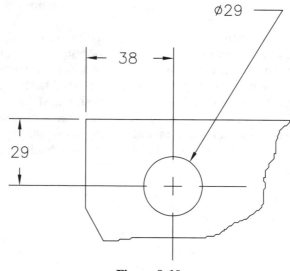

Figure 9-10

If no circular view exists for the hole, the leader line may reference the rectangular view of the hole. This often occurs in the case of objects which consist solely of cylindrical features. In this case, only a single profile view of the object may be required since the reader knows that all surfaces are cylindrical because of the included centerlines. When no circular view of the hole exists, the leader should point at the opening of the hole at the intersection between the centerline for the hole and its placement surface. This is shown in Figure 9-11.

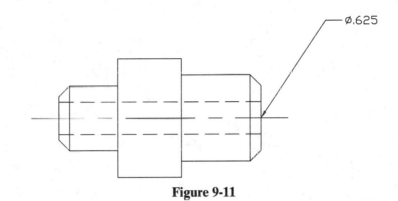

Figure 9-11

In this age of computer-aided drafting and design, many of these standards may be changing. For example, some software will not allow you to add the universal diametric symbol (Ø) to dimensions unless these dimensions are referencing an actual circle on the drawing. Thus, cylinders cannot easily be dimensioned with the diamteric symbol in the non-circular view as is standard dimensioning practice. Furthermore, different standards for the appearance of items such as arrowhead format and dimension values exist between the various engineering fields which make use of drawing documentation. In today's mechanical engineering field, fractional dimensions are rarely used. Rather, decimal dimensioning is preferred. To increase the readability of the dimension values, a system of unidirectional orientation is used. This means that the dimensional values are oriented such that they are read horizontally, from the bottom of the page.

This orientation has been shown in all of the dimensioning examples so far in this chapter and is depicted again in Figure 9-12. As also shown in this figure, the default precision for decimal inch dimensions in most mechanical engineering applications is two decimal places (hundredths).

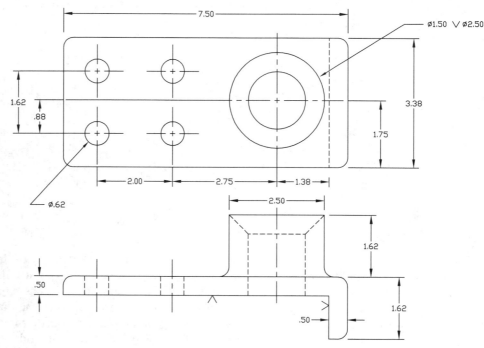

Figure 9-12

The use of a fixed decimal precision can cause some concern in the round-off of dimensional values. Many features of the mechanical engineering component are created through processes which use fixed size tools. An example of this would be a hole produced by a drill bit. These tools are still specified in terms of their fractional designation. The decimal equivalent of a 3/8 inch diameter drill bit is 0.375. Rounding in this case is difficult because of the '5' as the third digit. A simple rule of thumb may be followed in these cases; if the last digit is a 5, look to the previous digit. If the second to the last digit is odd, round up; if even, round down. In this way 0.375 would round to 0.38 and 0.125 would round to 0.12.

Other engineering fields may use a fractional format for dimension values. Fractional denominators are determined by subdividing successively by half. Thus, the fractions are in terms of halves, quarters, eighths, sixteenths, thirty-seconds, and so on. When a fractional format is used, the dimension values are usually placed in what is referred to as the aligned system. In this system, the dimensional values are aligned with respect to their dimension line and then read either from the bottom of the page or the right side. Figure 9-13(a) and (b) show examples of civil engineering drawings which use fractional dimensions in the aligned system. Further note the use of 'hash marks' in place of arrowheads in Figure 9-13(a), the alternative formats used for the 4 inch dimensions in Figure 9-13(b), and the use of local notes on both drawings.

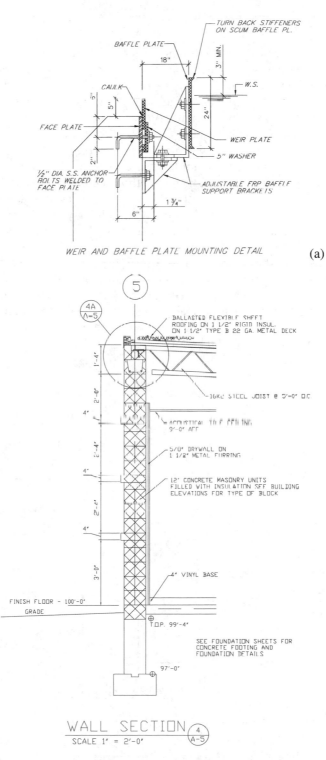

Figure 9-13 (a) Courtesy of Wade Trim/Associates. (b) Courtesy of Owen Ames Kimball Engineering

9.2 DIMENSION PLACEMENT

Many rules exist for the placement of dimension features in drawings. Possibly the single most important rule of dimension placement is that of using the most descriptive view. A dimension should always be placed in the drawing view where the feature being dimensioned is shown "in profile" or, in other words, where its shape description is most complete. As you become more adept at interpreting the graphic data given in orthographic projections, you will better appreciate this concept. The rationale of this rule is that the person reading the drawing should be able to interpret the dimension at its point of application. They should not have to search the drawing to identify what feature a dimension is referencing. A simple test of this technique is to examine a dimensioned drawing. Mentally cover views of the object such that only a single orthographic view is displayed. Examine the dimensions in the remaining view. Are you able to identify the shape of the feature being dimensioned without looking at the covered views? If so, the dimension is in an appropriate view. If not, then the covered view where you found the shape description is the proper choice. Figure 9-14 illustrates the concept of profile with regard to object features. You should note that more than one view may exist which provides an appropriate shape description for the feature being dimensioned. The best profile view is the one in which the feature is shown true shape and in object lines rather than in hidden lines. In the case of two equivalent choices, the attempt to obey other placement rules may determine the best location.

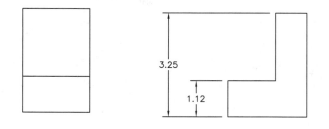

Dimensions reference
profile view of
feature

Dimensions do not
reference profile view
of feature.

Figure 9-14

The dimension format rules and the concept of dimensioning in profile should always be obeyed. Beyond that, many rules exist which aid in the placement of dimensions on a drawing. These rules should not be considered absolutes but merely preferences; that is, they may be broken in order to preserve the rules of format and profile. Although the lists of dimensioning rules may seem voluminous, they simply reflect a common sense approach to maintaining the readability of a drawing's dimensions. With experience, the use of the rules of placement will become obvious.

The following are some of the more common dimensioning rules of placement along with illustrations of each:

- The various dimension features should be created so that they do not interfere, obscure, or coincide with object, hidden, and center lines on the drawing.

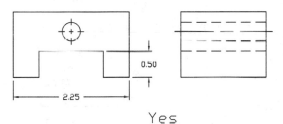

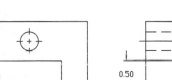

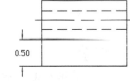

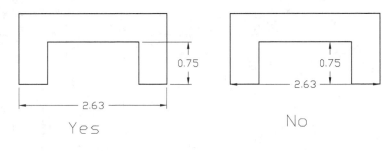

- Dimension lines should never coincide with object lines or other extension lines.

- Crossing of extension lines and dimension lines should be avoided if possible.

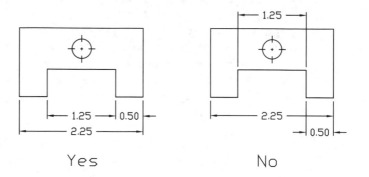

- Dimensions should reference object lines rather than hidden lines.

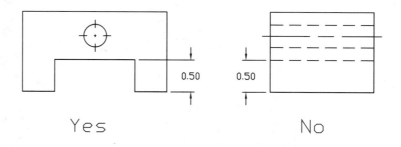

- Do not repeat dimensions on the same or another view.

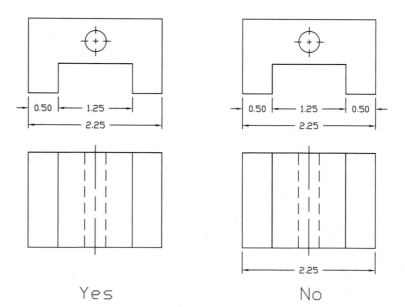

- Dimensions should be placed in clear spaces as close as possible to their point of application.

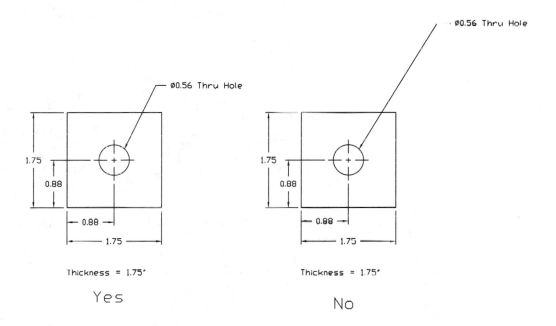

- Dimension and extension lines should plainly indicate their point of application.

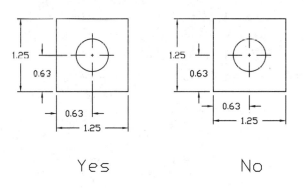

- Lettering should be distinct and clearly related to the proper dimension.

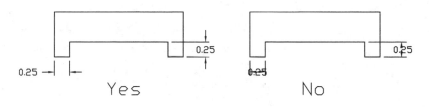

- No dimension should be placed such that its dimension line is closer than 3/8 inch from the object. Parallel dimension lines should be 1/4 inch apart.

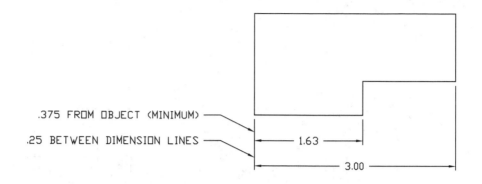

- When a 'chain' of linear dimensions is created, their dimension lines should be aligned.

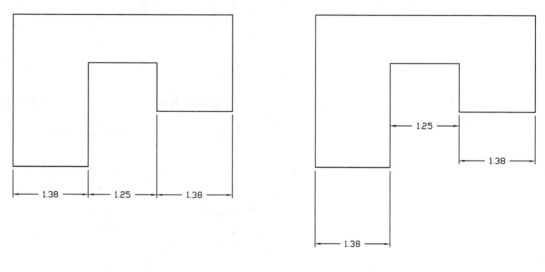

Yes   No

- When dimensions are 'nested,' the smaller dimensions should be placed closer to the object to avoid unnecessary line crossing.

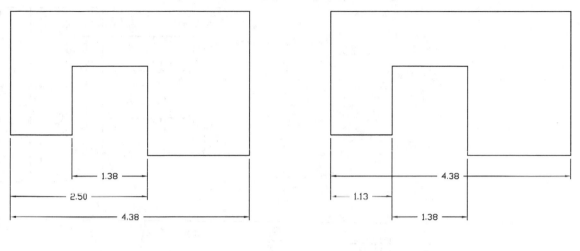

Yes   No

In addition to the local notes which reference specific features of the object such as holes, other annotation may be included on the drawing. Notes which refer to the entire object are referred to as general notes. These are placed in an open area on the drawing, normally in the center of the views and toward the bottom of the page. This will make the note more easily seen by those reading the drawing. A general note is used to specify that a drawing is in metric units. The word 'METRIC' is given in a box on the drawing. Examples of general notes are shown in Figure 9-15.

```
GENERAL NOTES:
ALL FILLETS AND ROUNDS .125R
REMOVE ALL SHARP CORNERS
ONE REQ'D, CAST ALUMINUM
HEAT TREAT TO ROCKWELL C42


GENERAL NOTES:
1. LANDSCAPING CONTRACTOR SHALL FINE GRADE AND
   SEED TO CONCRETE CURB.
2. EXTERIOR CONCRETE SURFACES SHALL HAVE COARSE
   BROOM FINISH.
3. ALL FILL WITHIN THE BUILDING AREA SHALL BE
   COMPACTED TO 95% OF MAXIMUM DENSITY AT OPTIMUM
   MOISTURE.
4. PROVIDE 6" pvc SLEEVES UNDER DRIVE AND ENTRY
   WALKS.
```

Figure 9-15

EXERCISES 9.2

In the following dimensioned drawing, a ballooned letter has been placed next to a dimension or group of dimensions. For each dimension or group, determine which rules of form and placement have been violated.

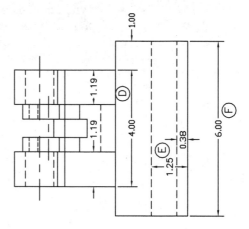

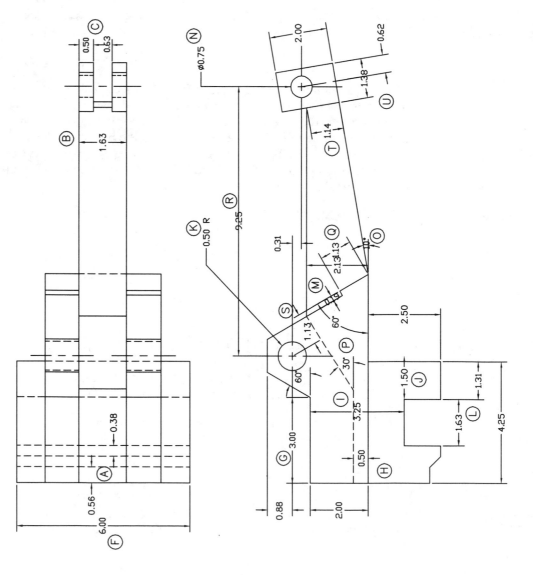

9.3 DIMENSIONING TECHNIQUE

In addition to knowing the language of dimensioning, it is important for you to understand which dimensions to include and how those decisions may be made. One simple approach to the basics of dimensioning practice is a process of decomposing a complex object into a collection of simpler geometries. Engineering graphics instructors frequently refer to this as the process of geometric breakdown. This means you will examine an object from the various shapes which comprise it and base your dimensioning scheme upon how those shapes are generated. Different geometric shapes require different dimensional parameters. For example, a block may be specified in terms of height, width, and depth and a cylindrical hole may be specified in terms of its diameter and its depth. Figure 9-16 shows several simple geometries and appropriate techniques by which they may be dimensioned.

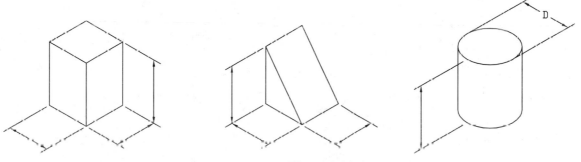

Figure 9-16

Figure 9-17 shows a block which has a smaller block removed from it. A dimensioning scheme for this object would include the overall dimensions of depth, width, and height for the large block (this is what we started with) as shown in Figure 9-17, and the depth, width, and height dimensions of the smaller block (this is the modification performed). A drawing for this is shown in Figure 9-18.

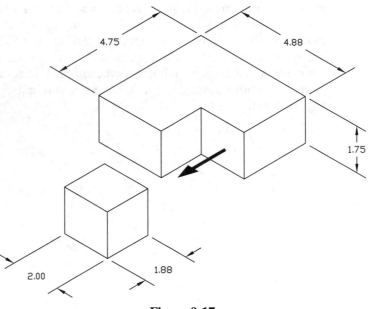

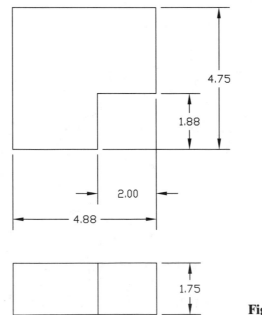

Figure 9-17

Figure 9-18

Dimensions are categorized as being either size dimensions or location dimensions. This is merely to indicate the function of the dimension. A size dimension gives the size of a feature, while a location dimension specifies the location of a feature of known size. Consider the example of a cylindrical hole. To create the feature on an object, you need to know the hole diameter and depth; these are the size dimensions. But you must also know where the location of the hole's center will be on the object. These specifications constitute the location dimensions for the feature. Figure 9-19 shows a hole feature added to the object shown in Figure 9-17. Note that the size dimension of diameter for the hole is given through the use of a local note (or callout).

Most modern computer-aided design software is dimensionally driven. This means that objects are easily modified according to the size and location of the features which make up the object geometry. For example, for the object shown in Figure 9-19, if the diameter (size) of the hole were to be modified, you would be able to merely select that dimension on the object and change the size of the feature. Likewise, the location of the feature can typically be controlled by dimensional constraints. You could reference the location of the center of the hole relative to the size of the block. In other words, you could say that the hole center be located at 30% of the length of a side of the block. Alternatively, you could say that you wanted the hole to always be located a given distance (say 1.55 inch) from one edge of the object. This type of dimensionally driven modeling makes design a flexible process where objects can be easily modified as needed.

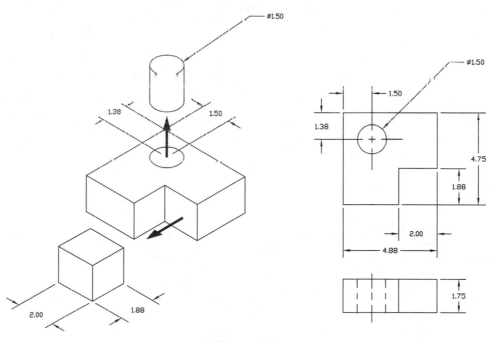

Figure 9-19

In general, we may simplify the process of applying geometric breakdown in dimensioning an object to a set of simple steps. These steps will not cover all possible situations but will give you a reasonable format by which to approach the majority of dimensioning situations.

1) Give overall dimensions except to the extents of cylindrical features. 2) For individual modifications to the original geometry, specify either the size of the simple geometry removed or the size of the features remaining after the removal (do not do both). 3) For cylindrical form features and holes, specify the size of the feature using the appropriate techniques and also locate the feature. Figure 9-20 shows an object dimensioned based upon this process.

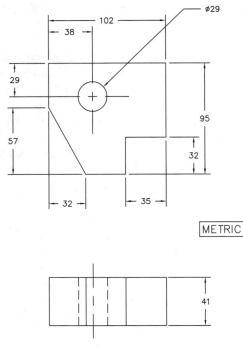

Figure 9-20

EXERCISES 9.3

For each of the following objects shown in Exercises 1–7, specify a full set of dimensions using the geometric breakdown technique. The dimensions are to be sketched freehand. Follow all the rules for dimension placement. No dimension values are required; however, include the R for radius and the diameter symbol where required.

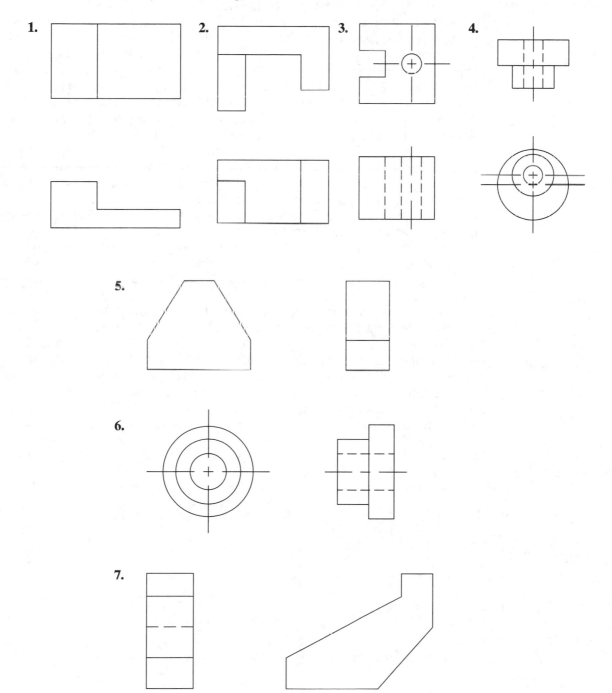

9.4 TOLERANCES

When you specify part dimensions, whether on a drawing or as part of the creation of a computer model file, you are defining a single unique geometry. However, because of the limitations of manufacturing, the parts produced from your specifications will include slight variations from this unique specification. These variations are not unexpected when you consider the possible areas for minute variances in creating the geometry. Slight differences in tool sizes, wear of the machine tools, human errors in both operation and measurement, among other factors all contribute to these variances. Furthermore, it is impossible to create something that has a dimension of exactly 2″–the limitations of the measurement tool may make it appear that a size is exactly 2″ when, in reality, the size is 2.000001″. While parts are inspected after fabrication to ensure compliance with the drawing or model specifications, the realities of economics force us to tolerate a certain amount of variation in size. Otherwise, more parts would be rejected than accepted and the costs of production would soar. By careful control of the acceptable manufacturing tolerances on components, interchangeability of mass produced parts may be achieved.

In practice, all dimensions in mechanical engineering include what is called a tolerance. Tolerance is the amount of acceptable dimensional deviation the designer will tolerate in the size of a feature. Parts are then inspected to ensure the actual size falls within the range defined by this tolerance. Tolerances may be specified in the title block of a drawing or within the setup of a model file (for computer-generated designs) and then this same tolerance is applied to all dimensions. This may be referred to as a global or general tolerance. A tolerance value may also be attached to the value of a specific dimension.

The dimensioned views shown in Figure 9-21 will be used to demonstrate the effect of tolerances in placing emphasis on features. The figure shows three identical views of a simple geometry. Note that only a single view orientation is being used to simplify the demonstration. Each of the views is dimensioned with a different set of dimensions. All three sets are valid with regard to form and placement. Each set of dimensions, however, place emphasis on different features of the object and hence if parts were produced through each set, all three parts would end up being slightly different. Let's assume the given dimensions are in inches and a global tolerance of ±.01 is specified. To observe the

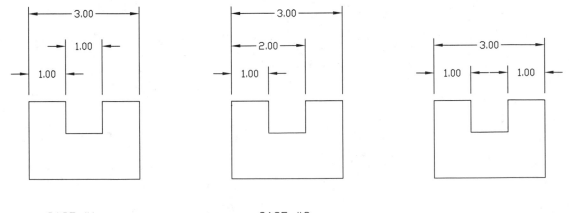

CASE #1   CASE #2   CASE #3

Figure 9-21

difference between the forms, examine the size of the notch opening for each of the dimensioned parts. In Case #1 the opening size is dimensioned directly, which means the size of the opening for a part which passes inspection can vary by only ±.01. In Case #2 the size of the opening is determined by the position of each of its side surfaces, each of which is dimensioned separately. Each of these two dimensions can vary by the specified tolerance and the part is still acceptable as dimensioned. In this case, since the size of the opening is dependent upon two dimensions its size can vary by ±.02 in a part which passes inspection. If you have trouble visualizing this, imagine the case in which the 1.00 dimension is at the low end of its acceptable tolerance (−.01) and the 2.00 dimension is at its high end (+.01). The dimensions are within tolerance and the slot is wider by 0.02. For Case #3, the size of the opening is still dependent upon the position of the side surfaces. However, in this case, each of these dimensions is being measured from a different side of the object and the position of the sides is dependent upon the third dimension. Now the size of the opening is dependent upon three dimensions and hence its size could vary by as much as ±.03 in a part which passes inspection as dimensioned.

A key consideration in the application of tolerances is that of mating dimensions. Mating dimensions refers to those features which must fit together when components are assembled. In the last example, you can easily imagine that some fashion of tab feature will be assembled into the slot. A potential feature is shown in Figure 9-22. When features on different parts are to be assembled, they must be dimensioned using the same technique. The tolerances on both mating parts then contribute to the process of assembly. If the format in Case #3 had been used to dimension the slot and the same dimensioning format as shown in Figure 9-22 were used for the tab, the following conditions for assembly would exist. Since the tolerances on each part contribute to the assembly, you should look at a worst-case situation. The worst case for assembly of the slot and the tab is when one feature is at its largest permissible size and the other feature is at its smallest. The largest permissible size for the tab which still passes inspection would be 1.03 and the smallest permissible size for the slot would be .97. Hence, in a worst-case situation, a total difference in size of .06 is possible.

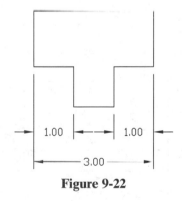

Figure 9-22

Perhaps the easiest area to see the effect of tolerances in dimensioning and the concept of mating dimensions is in the specification of location dimensions for hole patterns. When fasteners are being used to hold parts together, they must pass through or thread into these holes. The patterns, and especially the hole centerlines, must align in order for assembly to occur. Figure 9-23 shows a drawing of a cover plate with a rectangular pattern of holes. If this pattern is dimensioned using chain dimensioning techniques as shown in Case #1, accumulation of error occurs. For example, the distance between the centerlines of the hole labeled A and the hole labeled D is dependent upon three dimensions and their tolerances. If a tolerance of ±.01 is assumed, worst-case error would be ±.03. If the mating part is dimensioned in the same way, the potential misalignment of centerlines would be ±.06. As the number of holes increases, the number of dimensions upon which their placement is dependent increases. Hence, the size of the potential

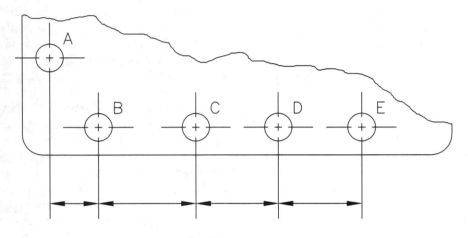

CASE #1

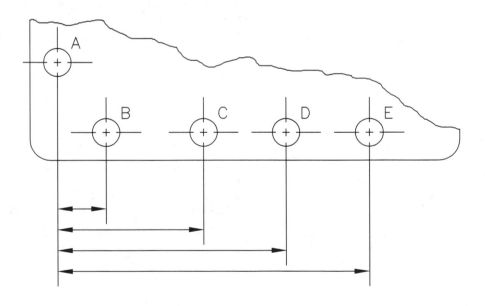

CASE #2

Figure 9-23

misalignment will increase. Case #2 shows the same pattern of holes dimensioned in a technique which is known as baseline dimensioning. In baseline dimensioning, a reference point or datum is chosen for each coordinate dimension. The features being dimensioned are then all referenced from this common datum (the baseline). When this technique is used to dimension a pattern of holes, the effects of error accumulation are minimized. Look at the dimensioned pattern in Case #2. The distance between the centerlines of any two holes is dependent only upon two dimensions at most.

Most hole patterns should be dimensioned in the following manner. 1) Choose a datum feature for locating the holes. Often, this is one of the holes in the pattern. Selection of a datum feature will be dependent upon how the parts assemble. Note that more than one datum may be required for some cases. 2) Ensure that the datum feature is located with respect to features of the part. 3) Locate all holes in the pattern from the datum. 4) Locate the holes in the pattern on the mating part in the same fashion using the same datum feature. Figure 9-24 shows three different patterns of holes that have been

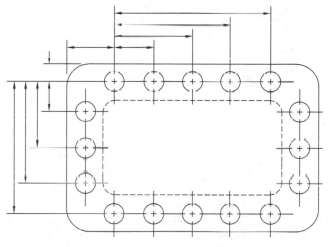

Example #1

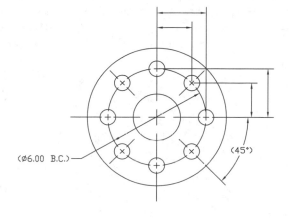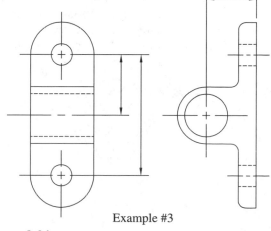

Example #2

Example #3

Figure 9-24

dimensioned. In each case, can you identify the datum feature? In the case of Example #3, one of the mounting holes is used rather than the larger hole which will likely accommodate some kind of a shaft. Remember that the same datum feature should be used for both patterns. With this dimensioning scheme, the mating part to which this component will be attached would have no indication of the location on the shaft. This is shown in the pictorial of Figure 9-25. Finally, notice the dimensions in Example #2. Two of the dimensions have been given in parentheses. A dimension in parentheses is referred to as a reference dimension. A reference dimension is a sort of "for your information" dimension. It is to be used to simplify the fabrication of the component but is not used for the pass/fail inspection of the part.

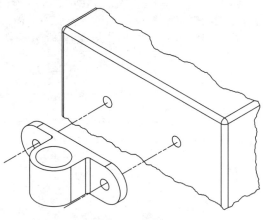

Figure 9-25

The exact or ideal value specified in the dimension is referred to as a basic value. This basic size is the value to which tolerances are applied. In addition, tolerances may be defined as unilateral or bilateral. This designation indicates the relationship between the tolerance value and the direction of its application (larger or smaller) to the basic size value. A unilateral tolerance value is applied only in one direction from the basic size, for example, 2.500 +.001 or 1.250 −.002. A bilateral tolerance value is one which is applied in both directions from the basic size. An example would be .8750 ±.0012. Note that the tolerance value is applied in each direction and therefore the total range of size variance is twice the tolerance value.

Tolerances applied to specific features may be written as a set of limits which define the range of the tolerance or as a set of ± dimensions. Figure 9-26 shows examples of tolerances written as limits and as ± dimensions. Notice from the example that ± dimensions may be written as unilateral or bilateral tolerances. Note also the form and arrangement of the dimension values as seen in this figure.

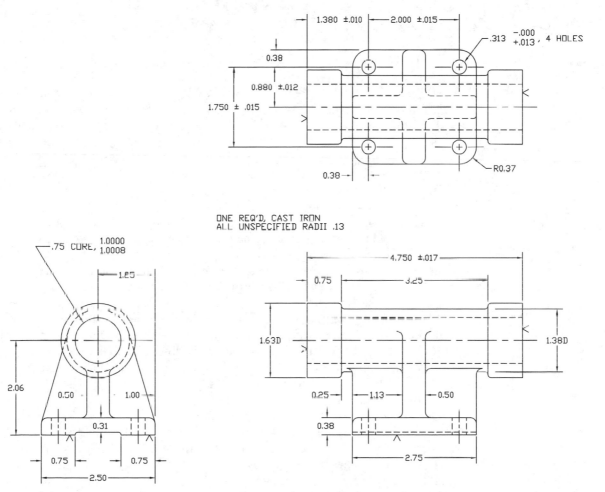

Figure 9-26

It is understood in mechanical design today that effective feature specification includes not just control of dimensional tolerances but also geometric relationships. This has given rise to the use of geometric tolerancing techniques. Geometric features such as flatness of a surface, straightness of an edge or axis, along with perpendicularity and parallelism relationships between surfaces and axes may be controlled through this technique. The technique referred to as Geometric Dimensioning and Tolerancing (GD & T) provides for the increased use of symbolic specification in dimensioning rather than text along with control of geometric features. A discussion of the principles of GD & T is beyond the scope of this text. The reader is advised to reference any one of the several specialized texts which discuss this technique if they wish to further their dimensioning skills.

Working Drawings

• •

As an engineer you will need to be able to interpret or read working drawings which have been created by others. There are many standards and conventions used in creating working drawings. These standards vary by industry and even within the same industry. In this chapter you will see several different types of actual engineering drawings.

10.1 READING CONSTRUCTION DRAWINGS

Drawings used in the design and construction of civil or architectural projects are frequently referred to as blueprints. The name *blueprint* is derived from an earlier era when construction drawings were reproduced by a method which resulted in a blue background with white lines and letters. Although modern-day construction drawings are made with dark lines on a white background, they are still often referred to as blueprints.

In the design and construction of civil projects, several drawings are necessary in order to completely describe the facility so that it can be built. The entire set of drawings is referred to as the "set of construction plans" or just "the plans." The specifications, or specs, consist of the written instructions regarding the construction of the facility. Together, the plans and specs make up the entire construction documentation. In this text, you will focus on understanding the construction drawings–specifications are beyond the scope of this text.

A set of plans usually consists of the following types of drawings: 1) site plan, 2) elevation views, 3) foundation plan, 4) floor plan, 5) sections, 6) detail drawings, and 7) other drawings as necessary for the construction of the project. Often, several drawings are required of each of these types of drawings for a complete set of plans. Some of the additional drawings which may be required depend on the specific project and may include the following: roof plan, floor joist plan, electrical plan, heating and ventilation plan, plumbing plan, and a framing plan. In addition to these drawings, schedules for materials which list the type of doors and windows to be used in the construction of the facility are also usually included with the plans. For road construction projects, a set of drawings usually consists of site maps, plan and profile drawings, sections, and detail drawings.

In civil applications, plan views are views which are made from a vantage point *above* the "object." Thus, plan views can be thought of as *top* views. You are probably familiar with the term **floor plan**. A floor plan is a drawing made from a vantage point above a building which shows the layout of all of the rooms on a particular floor. A foundation plan shows the building foundation from above, the electrical plan shows the wiring diagram for the building from above, the heating and ventilation plan shows the location of the ducts and equipment from above, and so on. Profile views are any views which show the building or project from the front or the side. In other words, they are views where the top of the structure is seen as an edge.

In the construction industry, the orientation and location of the project with respect to its surroundings is extremely important. Imagine the problems which would develop if a building were constructed on someone else's property. There are several methods used on drawings to help the contractor locate the structure properly. Bearings of lines are usually shown on the drawing. Recall that the bearing of a line is the angle that the line makes with a North–South line, as illustrated in Figure 10-1. Bearings of lines are only seen in plan views, i.e., from above. On the construction site, bearings of lines can be obtained by any of several surveying techniques and the building can then be accurately located on the property.

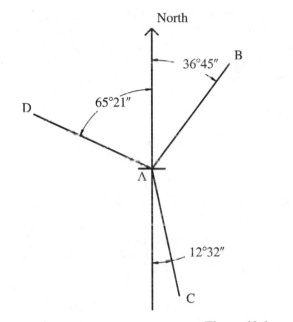

Bearing AB = N 36° 45″ E

Bearing AC = S 12° 32″ E

Bearing AD = N 65° 21″ W

Bearing BA = S 36° 45″ W

Bearing CA = N 12° 32″ W

Bearing DA = S 65° 21″ E

Figure 10-1

Control points are also often given on construction drawings in order to accurately locate features of the project. By this method, an "origin" for the construction site is designated and all points are referenced as north, south, east, or west from it. Thus, a point on a drawing might have coordinates such as N 13750 and E 7895. Similar to bearings, the coordinates of the points on the drawing are only seen in plan views. The origin for the system is usually referred to as a bench mark. Bench marks have been established across the nation by the United States Geological Survey (USGS) and typically consist of a concrete cylinder imbedded in the earth with a brass circular medal on top. The location of each bench mark in the country has been determined with a high degree of accuracy. Many times, job bench marks are established on construction sites if a USGS bench mark is not located within the vicinity of the project.

Bench marks are also used to determine the elevations of points on the construction site. Elevations are used to establish vertical distances between points on a building. For example, the top of a floor slab might be specified as having an elevation of 556 feet. Elevations will only be seen in profile views and are usually either referenced to true elevations, i.e., the height of the point above sea level or to job elevations. With job elevations, a bench mark is established and given an arbitrary elevation of something like 100 feet. All other elevations on the plan are then specified relative to this point. A bench mark elevation of 0 is usually not specified to ensure that job elevations are never negative.

Since the drawings for a project usually consist of several sheets, the first part of the set of plans contains an index of all of the drawings in the set. This is usually followed by a glossary of terms found on the drawings. For example, the term ELEV may be designated as the abbreviation for *elevation*.

Site Plan

The first drawing or group of drawings in a set of construction plans is a site map or a site plan. Figure 10-2 shows a typical site plan for a construction project. Note that one of the project base lines shown on this drawing has a bearing of N 37° 34′ 09″ E and that several control points are shown on the drawing. For example, the control point at the center of the lower right circular structure has coordinates of N345275.3 and E426932.8. Note also that, in the table shown on the drawing, the locations of points 509 and 510 are specified (point 510 is a railroad spike). This particular site plan is superimposed on a topographic map of the area, although this is not always the case.

Elevation Views

Elevation views show the project from the outside as it would look when completed. The elevation views show the various "elevations" of the project and you can think of them as similar to front or side views. Typically, elevation views are labeled as either North, South, East, or West (however, they are sometimes labeled as Front, Rear, Right, and Left). A North Elevation would show what the completed building would look like if you stood to the north and looked back at the building, a South Elevation shows what it looks like from the south, and so on. Although elevation views do not contain a lot of details and dimensions about the actual construction of the building, they do help contractors and owners visualize the resulting project before construction takes place. Figure 10-3 shows four elevation views of a building to be constructed as a part of a wastewater treatment facility.

Foundation Plan

The remainder of the drawings in the set of construction plans are typically included in the order in which the building is constructed. Because a building is constructed from the foundation up, the foundation plans are among the first drawings in the set. A building foundation is almost always constructed out of concrete which has been reinforced with steel bars or rebars (<u>re</u>inforcing <u>bar</u>). Concrete footings support the walls and columns in a building and the foundation walls themselves are also many times made out of reinforced concrete. Details about the size of the footings and the size and location of reinforcing bars are usually included in a wall sectional view. Sometimes a reinforced concrete slab is also constructed for the building and is included as a part of the foundation plan or in a wall sectional view. Concrete slabs will typically contain reinforcing bars as well as a steel mesh for control of the thermal expansion and contraction of the slab.

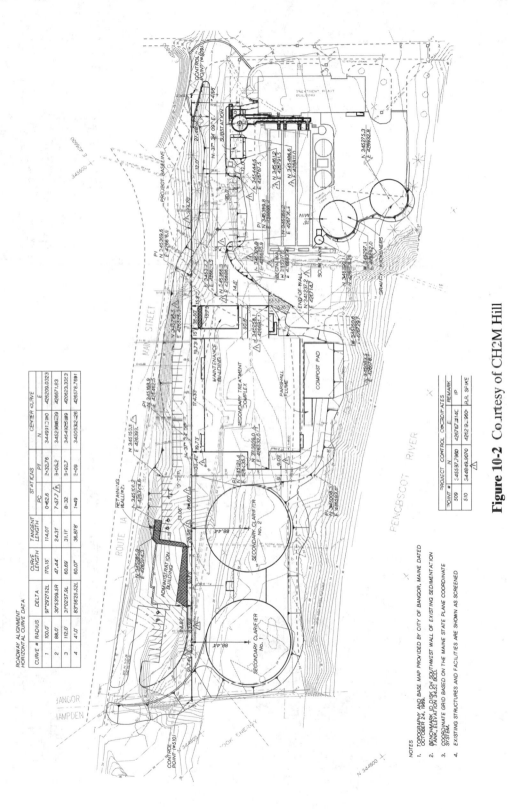

Figure 10-2 Courtesy of CH2M Hill

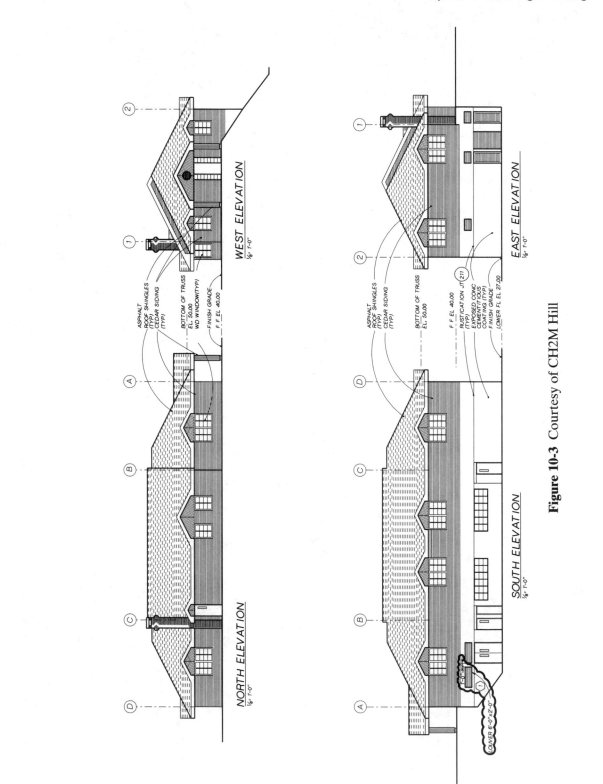

Figure 10-3 Courtesy of CH2M Hill

Figure 10-4 shows a foundation plan from a building project. In this drawing, the right side of the foundation is a "slab" which has overall dimensions of 37′ 8″ wide by 26′ 0″ wide. The reinforcing bars in the walls are shown throughout, i.e., "A" BAR #6@12″. This means that the rebars are #6 (3/4″ diameter) and are located at a distance of 12″ center-to-center. On the left side of the structure, the outline of the foundation is shown as a solid line on the outside of the building and as a dashed line on the "inside" of the build-

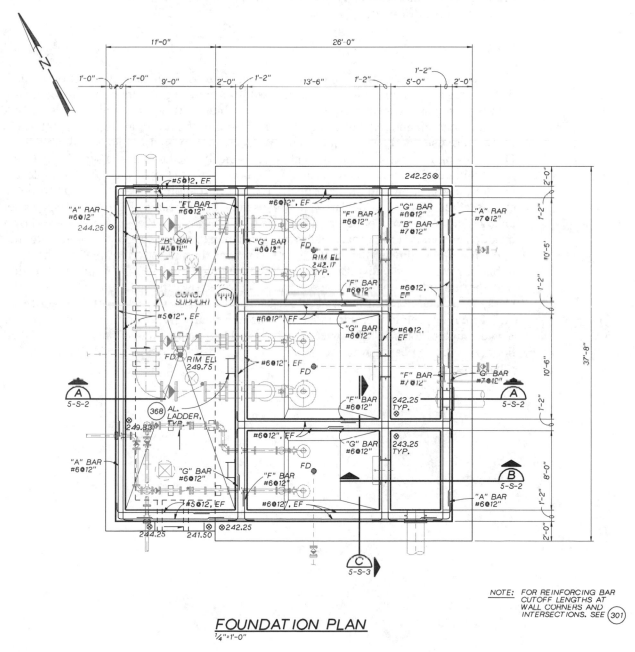

FOUNDATION PLAN
¼″=1′-0″

Figure 10-4 Courtesy of CH2M Hill

ing. The footing on this part of the building is only 3′ wide and is located only under the structure walls. Note also that floor drains (FD) are specified in the center of the three middle sections of the building. Three sections are "cut" through the building: A–A, B–B, and C–C. Note that these are referenced to other sheets within the set of plans. For example, section A–A will be found on sheet *5-S-2* of the plans. Another feature of this particular foundation plan is that the elevations at the top of the slabs/footings are shown in the drawing. For example, the slab on the right has an elevation of 242.25′ (denoted as ⊗242.25 on the drawing).

Floor Plans

The type of construction drawing you are probably most familiar with is a floor plan. A floor plan shows the layout of the rooms within a building. Doors between rooms are shown as well as the location of windows, closets, plumbing fixtures, and any other pertinent information about the drawing. The dimensions of the rooms as well as the thickness of walls are usually shown on floor plans. Walls are shown on floor plans as double lines. One drawing for each floor in the building is typically included in the set of plans. Sectional and detail views are also usually referenced from the floor plan. For example, for an interior wall shown on the floor plan, a reference to another drawing which shows the cross-section of that particular wall will usually be indicated. Figure 10-5 shows an enlarged floor plan of a building. The rooms from this partial floor plan which are shown include restrooms, a cafeteria, a hallway, a mechanical room, and a coat closet. Note that in the restrooms, the location of the toilets as well as the sinks are shown. Since these are public restrooms, the individual stalls are shown along with dimensions and the direction that the individual stall doors open. The cafeteria shows the location of the sink and refrigerator. The non-bearing walls are designated with the note: 8″ CMU UP TO 10′-0″ AFF. The interpretation of this note is that the non-bearing walls are to be constructed of 8″ wide cinder blocks (CMU-concrete masonry unit) to a height of 10′ above the finished floor (AFF). The bearing walls of the building are to be constructed of 12″ CMU with 2 #5 rebars located 24″ on center.

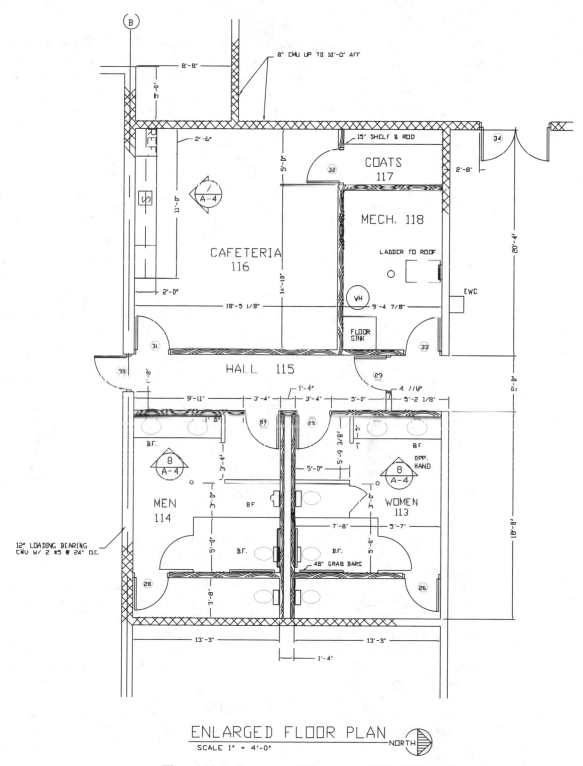

ENLARGED FLOOR PLAN
SCALE 1" = 4'-0"

NORTH

Figure 10-5 Courtesy of Owen Ames Kimball Engineering

Sections

Sections in construction drawings can be broken down into two types: general sections, which show room or floor layouts for buildings, or detailed section views which show cross-sections with enough detail for construction purposes. In fact, a floor plan could also be thought of as a horizontal section through a building. Figure 10-6 shows a vertical section through a house. Note that with this type of general section, not enough detail is included for construction purposes, but it is helpful because it gives you a general idea about the layout of the rooms and floors within the house.

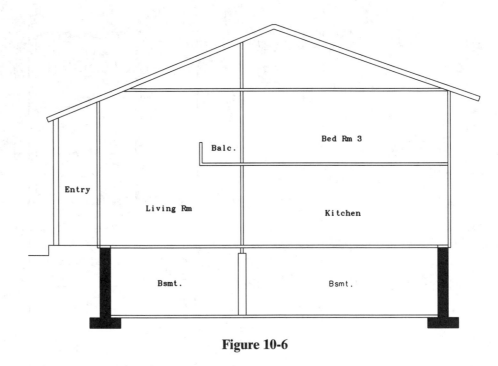

Figure 10-6

Detail sections show a great deal of information. They show how the different components in a building system fit together and they provide information which cannot be shown in large-scale drawings like floor plans or elevation views. Wall sections are among the most prevalent type of sectional drawing although roof framing and foundation sections are also common. Figure 10-7 shows a section A-A for the building whose foundation plan was shown in Figure 10-4. Note that the various materials to be used in the foundation/wall system are all shown in this sectional view (for example, the granular and concrete fill and the location and size of the rebars). Some of the rebars which are located in the foundation walls have designations of EW (Each Way) and EF (Each Face). Note also that the foundation to the right is a "slab" type of foundation, whereas the footing on the left is only 3'-0" wide as discussed previously. The elevations of the top of the slabs and the foundations are shown in this section as well as the location of pipes, a ladder, and a handrail.

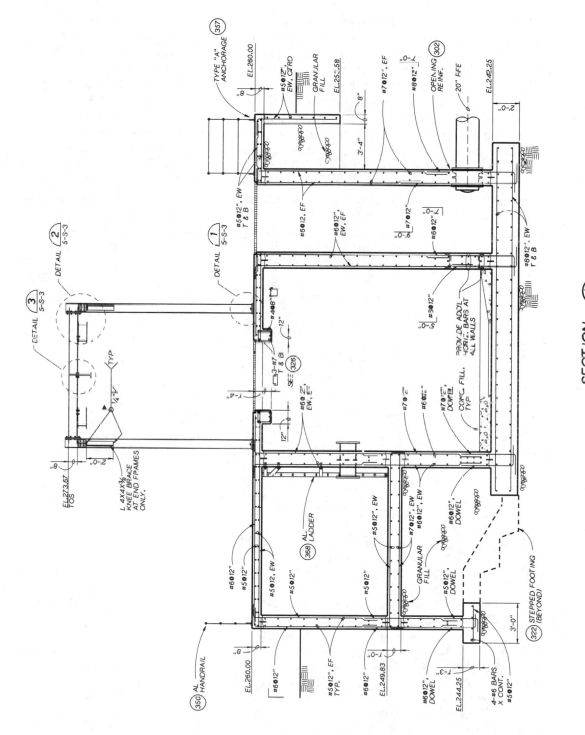

Figure 10-7 Courtesy of CH2M Hill

Details

Detail drawings are made to show one or two particular features on the constructed facility so that it can be built. Because constructed facilities are typically large, some of the finer details of the construction cannot adequately be shown on other types of drawings. Thus, detail views show one specific area on a drawing which has been "enlarged." Detail views are typically shown on the same sheet as the elevation view or floor plan. If they are not found there, they will be referenced on the elevation view or floor plan and included on a different sheet. Sometimes, all construction details are shown on several detail sheets. Figure 10-8 shows a detail drawing for a column–beam connection in a particular building. In this detail, the members are labeled (e.g., W8X18 is a certain size of structural steel I-beam). Welds are called out on this drawing as well as the bolts to be used in the connection (8- 7/8″ diameter A325 High Strength bolts). The plates used in the connection are also specified; for example, at the column–beam interface, a plate which has the dimensions 7/8″X9″X1′2″ is to be used in the connection.

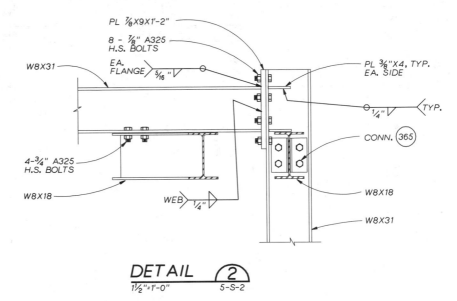

Figure 10-8 Courtesy of CH2M Hill

Plan and Profile

Another common type of civil engineering drawing is a plan and profile drawing. Recall that plan views show a structure from above. In other words, plan views show differences in *bearings* of the lines which define the structure. Profile views show the structure from the side or the front. Thus, profile views show changes in *elevation* for the structure. Figure 10-9 shows a plan and profile drawing for a simple structure (a retaining wall). Note that in this drawing the plan view shows the changes in bearing for the wall, and the profile view shows the change in height of the wall. Note also the conventional practice of showing the profile view in an aligned position, i.e., the left portion of the plan view of the wall has been revolved so that it is parallel to the profile view before projecting it into that view. This is in keeping with conventional practice as described in Chapter 6.

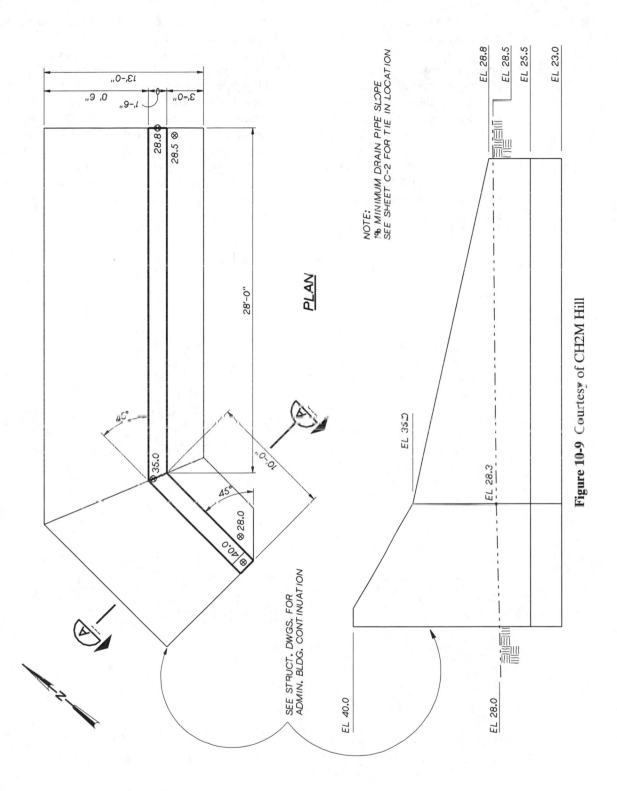

Figure 10-9 Courtesy of CH2M Hill

Perhaps the most common use for plan and profile drawings is in the field of road construction. Thus, changes in bearing (horizontal curves) are shown in the plan view and changes in elevation (vertical curves) are shown in the profile view. Horizontal curves occur when you are traveling on level ground and go around a curve in the road. Vertical curves occur when you are going up or down hills on a given road. Figure 10-10 shows a plan and profile drawing for a section of roadway which is to be improved. Note that in this drawing the profile of the road centerline is shown as well as the profiles of the east and west top of curb (T/C) for the road. As you can visualize from the profile drawings, if you were driving along this road, the railroad tracks are located at the peak of a hill. Also shown are the changes in elevations for the east and west right of way for the road as well as the existing profile at the centerline of the road. At the location of the railroad tracks, there is a significant "bump" in the existing pavement which should be smoothed out after the road is improved.

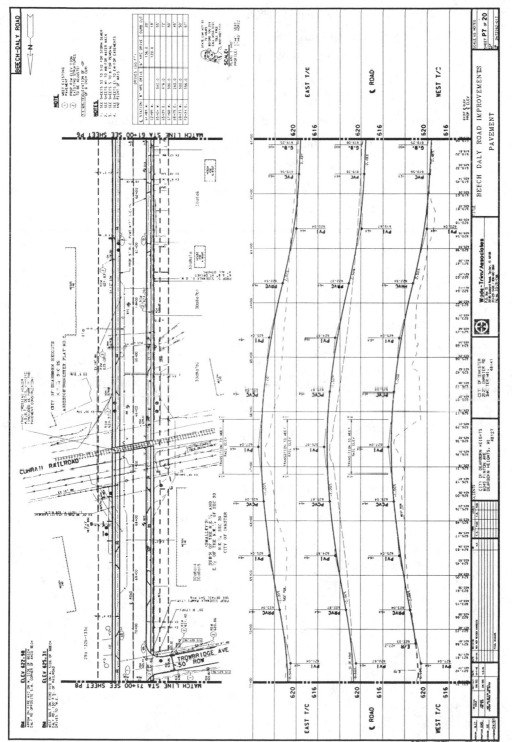

Figure 10-10 Courtesy of Wade Trim/Associates

EXERCISES 10.1

1. Answer the following questions about the section and plan drawings shown in Figure 10-11.

   What material is specified in the handrail for the stairway?_____

   What material is specified in the handrail for the extension?_____

   For a typical stair, what is the height of the riser?_____

   What is the depth of a tread?_____

   What is the total run of the treads?_____ of the risers?_____

   What is the run of the landing?_____

   What is the size of the toeplate?_____

   What is the material used in the landing?_____

   What is the material used in the stairtread?_____

   What is the distance between the top of the handrail and the top of the stringer?

   _____

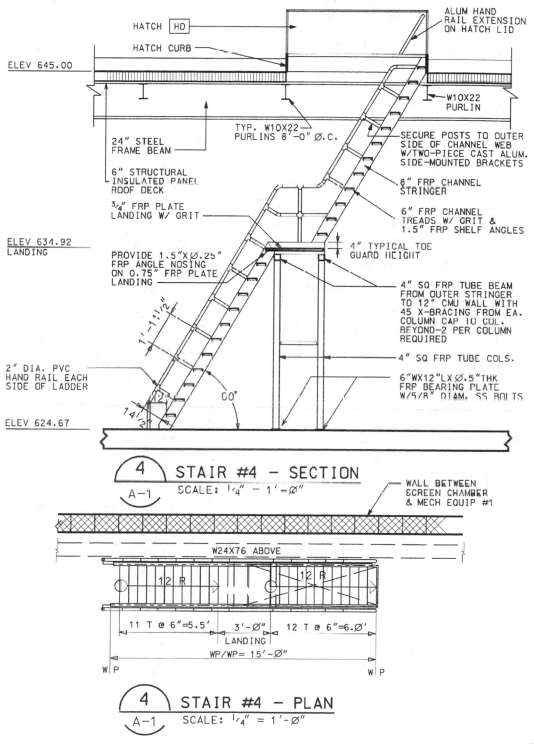

ALUM HAND RAIL EXTENSION ON HATCH LID

HATCH HD

HATCH CURB

ELEV 645.00

W10X22 PURLIN

TYP. W10X22 PURLINS 8'-0" Ø.C.

SECURE POSTS TO OUTER SIDE OF CHANNEL WEB W/TWO-PIECE CAST ALUM. SIDE-MOUNTED BRACKETS

24" STEEL FRAME BEAM

8" FRP CHANNEL STRINGER

6" STRUCTURAL INSULATED PANEL ROOF DECK

6" FRP CHANNEL TREADS W/ GRIT & 1.5" FRP SHELF ANGLES

3/4" FRP PLATE LANDING W/ GRIT

ELEV 634.92 LANDING

4" TYPICAL TOE GUARD HEIGHT

PROVIDE 1.5"X Ø.25" FRP ANGLE NOSING ON 0.75" FRP PLATE LANDING

4" SQ FRP TUBE BEAM FROM OUTER STRINGER TO 12" CMU WALL WITH 45 X-BRACING FROM EA. COLUMN CAP TO COL. BEYOND-2 PER COLUMN REQUIRED

1'-11 1/2"

4" SQ FRP TUBE COLS.

2" DIA. PVC HAND RAIL EACH SIDE OF LADDER

6"WX12"LX Ø.5"THK FRP BEARING PLATE W/5/8" DIAM. SS BOLTS

60°

ELEV 624.67

4 STAIR #4 - SECTION
A-1   SCALE: 1/4" = 1'-Ø"

WALL BETWEEN SCREEN CHAMBER & MECH EQUIP #1

W24X76 ABOVE

12 R   12 R

11 T @ 6"=5.5'   3'-Ø"   12 T @ 6"=6.Ø'
                 LANDING
         WP/WP= 15'-Ø"
W P   W P

4 STAIR #4 - PLAN
A-1   SCALE: 1/4" = 1'-Ø"

Figure 10-11 Courtesy of Wade Trim/Associates

2. Answer the following questions about the drawing shown in Figure 10-12. Note that this figure represents a sectional view of a *cylindrical* water tank.

What are the following elevations?

   Low Water_____

   High Water_____

   Flood Stage _____

What is the thickness of the concrete footing? _____

What size rebars are used in the footing?_____

What spacing are they center-to-center?_____

What size rebars are used in the upper slab?_____

What spacing are they center-to-center?_____

How far away from the wall is the pipe support located?_____

How many steps are there in the wet well?_____

What is their spacing?_____

What is the overall height of the structure?_____

What is the diameter of the wet well?_____

What is the diameter of the concrete slab/foundation?_____

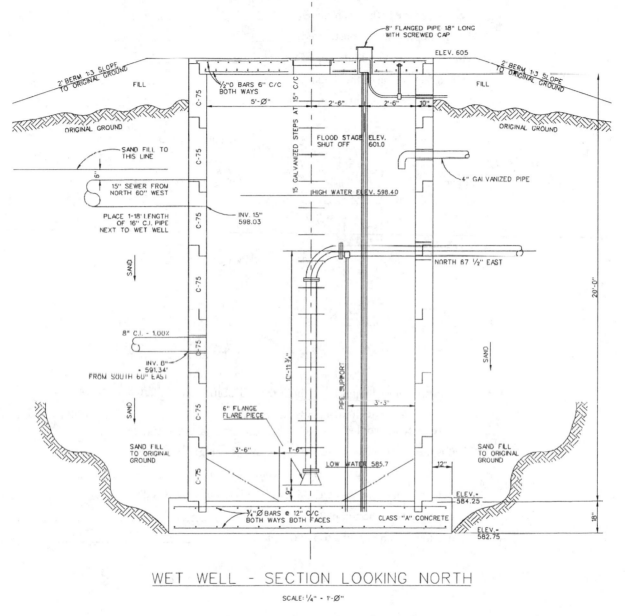

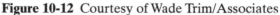

WET WELL - SECTION LOOKING NORTH

SCALE: 1/4" = 1'-0"

Figure 10-12 Courtesy of Wade Trim/Associates

3. Answer the following questions about the shutter window detail shown below.

What is the size of the deep shelf?_____

How is the shelf to be constructed?_____

What is the elevation at the top of the shelf?_____

What is the spacing of the metal studs used in the wall?_____

What is the size of the overhang on the left side of the wall?_____

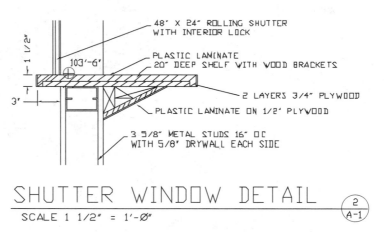

SHUTTER WINDOW DETAIL
SCALE 1 1/2″ = 1′-Ø″

Courtesy of Owen Ames Kimball Engineering

4. Answer the following questions about the wall section shown in Figure 10-13.

On which other drawing sheets would you find details that show how the wall/beam connections are to be accomplished?_____

What is the elevation at the top of the column footing?_____

What is the elevation at the top of the concrete slab?_____

How and of what materials are the lower walls to be constructed?_____

What is the size of the top plate?_____

What is the elevation at the top of the plate?_____

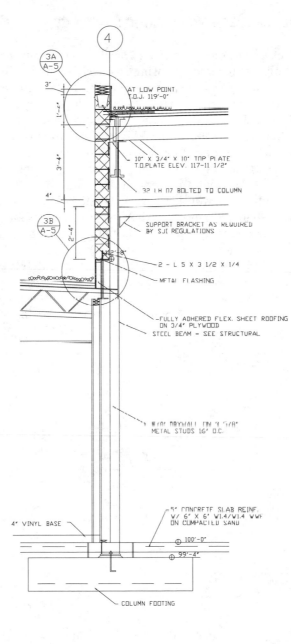

Figure 10-13 Courtesy of Owen Ames Kimball Engineering

10.2 *MECHANICAL ENGINEERING WORKING DRAWINGS*

The evolution of design and production of a mechanical part or system involves an entire family of technical drawings. Collectivly these drawings are often referred to as working drawings. Working drawings provide documentation for the sequence of design and fabrication. In a traditional mechanical design sequence, this family would include a design layout, a set of detail drawings, and an assembly drawing complete with a parts list and often a bill of materials. While the increased presence of CAD systems in the workplace have helped to drive a move toward concurrent rather than sequential design techniques, an examination of this family of working drawings is still valid.

All working drawings will include some form of annotation. The one annotation found on all working drawings is the titleblock. The titleblock is located in the lower right-hand corner of the drawing and includes information such as the name and address of the company who created the drawing, the name or title of the part or assembly, a drawing number for administrative purposes, and specification of the drawing scale. Figure 10-14 shows the standard format for titleblocks as specified by ANSI. Some companies may have developed their own format to include or omit certain information as their needs require. Other annotation found on working drawings may be specific to the type of working drawing. Some drawings include dimensions and notations, others do not. As we examine the common types of working drawings, we will note these differences.

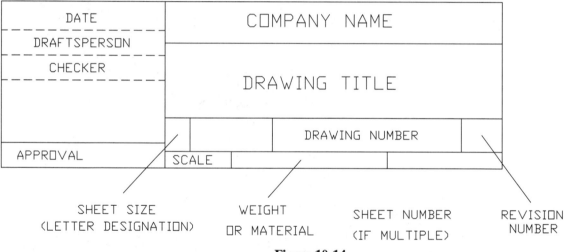

Figure 10-14

Finally, it should be remembered that engineering drawings constitute a legal document. All efforts must be made to ensure that the information included in the drawing, whether graphical or annotative, is accurate and unambiguous.

The Design Layout

Traditionally the design layout is the domain of the design engineer. After the development of preliminary ideas, which are often simply sketched, the engineers put their concepts to paper in the form of a layout. A layout is a set of orthographic views showing the mechanism or system in an assembled format. The function of the design layout is to show the relative sizes and features of the parts making up the system, how these parts fit together, and whatever motions occur in the function of the system. The layout is created full scale (if practical) and the only dimensions and annotation included are those items which could not be taken directly (by taking measurements of the drawing or querying the CAD database) from the geometry of the represented parts. Size and location dimensions of part features may be taken directly from the geometric representation of the part regardless of whether the layout is created on paper or electronically. Examples of annotation which cannot be directly taken from the part geometry only include items such as the range of motion a member may move through, required surface finishes and treatments, and those tolerances the engineer feels are critical to the design.

The design layout is used to check for interferences between components, both in their assembly and during any motions which occur during the operation of the system. It is used to determine which features and surfaces are critical to assembly and alignment. In general, layout drawings are constructed with all the information required for the creation of the next level of working drawings: the detail drawing.

The move to a computerized workplace has changed the face of design representation, documentation, and analysis. In particular, the recent development of dimensionally driven, feature-based solid modeling systems has provided the engineer with tools which can replace the traditional design layout. These packages allow the engineer to use the modeling system as a tool for design development rather than simply as a means for design documentation. Individual part models are created for each of the components in the assembly.

In addition, some current professional level solid modeling systems include a "layout mode." This function enables the user to define basic relationships and positioning information for the components of an assembly without having to specify detailed part geometries. These computer "layouts" are typically not precision scaled models but rather 2D representations intended to show the position and size envelope for each component. The electronic "layout" may then be associated within the CAD system with the detailed 3D solid model part files as they are created. In this way, the layout and the part files may be used to create automatic assemblies.

Detail Drawings

Once completed, the design layout is turned over to a group of draftspersons known as detailers. It is the responsibility of the detailers to create the required detail drawings for the mechanism components. In the world of solid-modeling CAD systems, these detail drawings may be extracted directly from the solid model geometric database. Probably the most common of the working drawings, the detail drawing is (typically) a multiview drawing of a single part, complete with all notes, dimensions, and tolerances. Detail drawings will include sections and auxiliary views as required to fully describe the geometry. The purpose of the detail drawing is to provide a complete engineering definition of the component. It will serve as a means for part inspection and quality control. The detail drawing must include all the information required for fabrication of the component. It should provide sufficient information for the creation of fixtures and toolings for the part's manufacture, and for manufacturing process planning. However, in modern practice the detail drawing will not normally include specific manufacturing or machining instructions.

A detail drawing is required for every component in the assembly which is to be manufactured. Many companies will contract with other firms for the manufacture of some or all of the components of an assembly. These contracted companies must be provided complete and unambiguous specifications for the parts they will produce.

Figures 10-15, 10-16, and 10-17 depict various detail drawings of bicycle components. For example, Figure 10-15 shows a detail drawing for a Brake Bridge. Note that in addition to the principle views shown (in this case a left-side, front, and right-side view), the drawing also includes a scaled view of a part detail (lower left corner) and two section views.

Figure 10-16 is a detail drawing of the Seat Lug for a bicycle. This detail drawing is interesting in how the dimension values for the various features are specified. The dimensions do not include values but rather numbers which reference columns in the table at the upper right. The dimension values are then given in the table. This technique was used to minimize clutter in the drawing and to enhance readability of the dimensions.

The detail drawing will usually include a table of drawing revisions. These revisions to the drawing can include changes to specific dimensions and specifications or the addition of new features to the part. The revision entries are always dated. The revision table is typically placed in the lower right-hand corner on the drawing sheet above the title block. Examples of this list of revisions may be seen in the detail drawings of Figures 10-15, 10-16, and the tabular detail drawing of Figure 10-17.

Tabular drawings are a special type of detail drawing. Tabular drawings are used to document a series of parts which have similar shapes but different dimensions. These are referred to as part families. The tabular drawing consists of a set of orthographic views which depict a nominal part representation. The part is dimensioned completely but the values of the dimensions are given as letters rather than values. The letters correspond to entries in a table of dimension values, shown in this case in the upper left-hand corner of the drawing. In this way, several parts may be specified through a single drawing. Figure 10-17 shows a good example of a tabular drawing. The drawing is of a set of handlebars for a bicycle. Because the same basic design of handlebars will be used on different bicycles, a tabular drawing was used to specify the family of parts. Notice that not all of the dimensions specified in the drawing must be dimensioned through the table. Further notice that tolerance values for the respective dimensions are also given in the table.

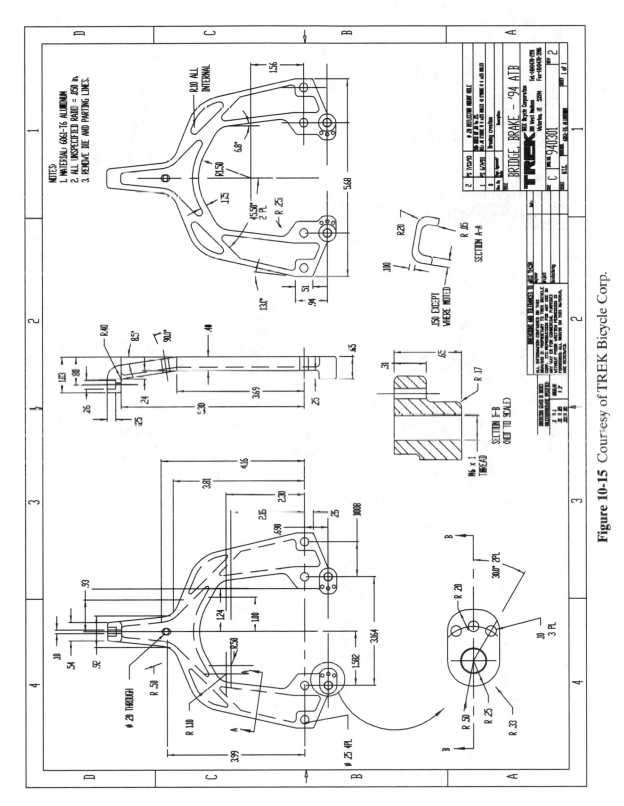

Figure 10-15 Courtesy of TREK Bicycle Corp.

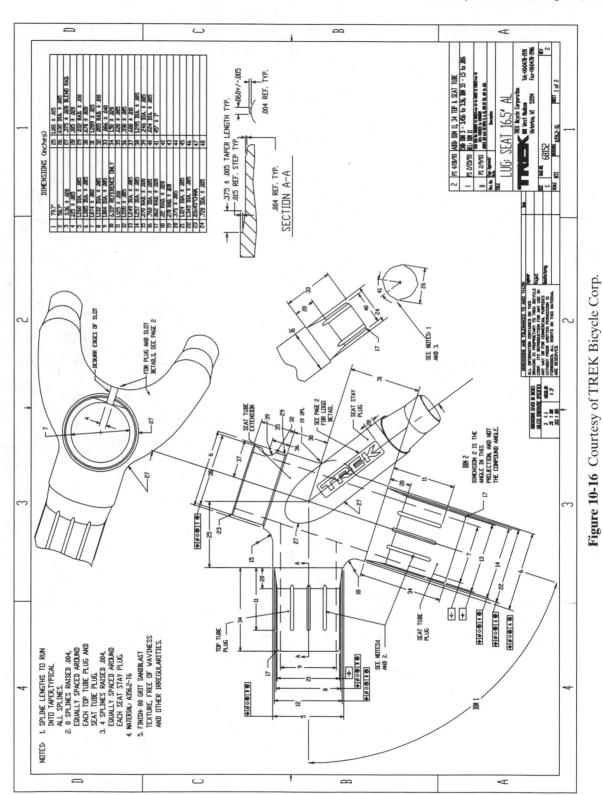

Figure 10-16 Courtesy of TREK Bicycle Corp.

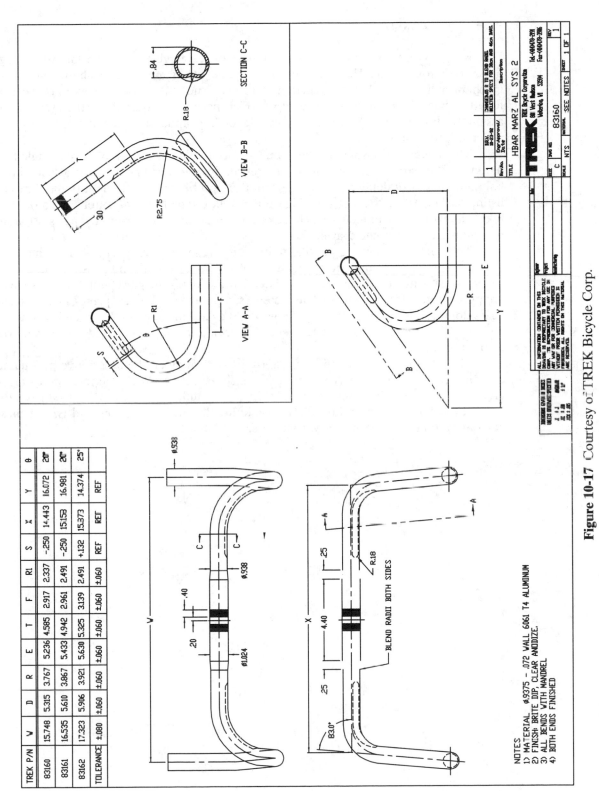

TREK P/N	V	D	R	E	T	F	R1	S	X	Y	θ
83160	15.748	5.315	3.767	5.236	4.585	2.917	2.337	-.250	14.443	16.072	20°
83161	16.535	5.610	3.867	5.433	4.942	2.961	2.491	-.250	15.153	16.981	20°
83162	17.323	5.906	3.921	5.630	5.325	3.139	2.491	+.132	15.373	14.374	25°
TOLERANCE	±.080	±.060	±.060	±.060	±.060	±.060	±.060	REF	REF	REF	

NOTES:
1) MATERIAL ⌀.9375 – .072 WALL 6061 T4 ALUMINUM
2) FINISH: BRITE DIP, CLEAR ANODIZE.
3) ALL BENDS WITH MANDREL
4) BOTH ENDS FINISHED

Figure 10-17 Courtesy of TREK Bicycle Corp.

Assembly Drawings

The assembly drawing represents the epitome of the engineering drawing as a design documentation device. The assembly drawing, as the name implies, shows the entire mechanism as assembled. Its function is to show not only how the various parts fit together but to also provide a record of which parts are required for the system. The assembly drawing can take many forms depending upon the specific application for which it is to be used.

The assembly drawing may be a multi-view drawing or consist of only a single profile view. The parts of the system are referenced by a leader line which is attached to a ballooned letter or detail number. This letter or number is used to identify the part in a listing of the parts in the assembly. Figure 10-18 shows the frame assembly of a mountain bike. The various components which make up the assembly are given in a parts list along the right-hand side of the drawing.

The parts list may also provide information regarding the part material, the minimum number of each component required for one full assembly, and possibly a reference to the drawing number of the associated detail drawing for the part. Standard parts which are purchased "off-the-shelf" such as fasteners, retaining rings, pins, springs, bearings, and gears do not require detail drawings, however, they are included in the assembly drawing and the parts list. Such parts are referenced as "standard" or "purchased" in the material column of the parts list.

The parts list in Figure 10-18 specifies the part number, the descriptive part name, and the quantity of the part required for each assembly (in parentheses and only if greater than "one"). For example, detail #16 is part number 950113, is referred to as the Cable Stop Guide (supports the shifter cable for the rear derailleur), and two of these parts are required for each bicycle frame assembly.

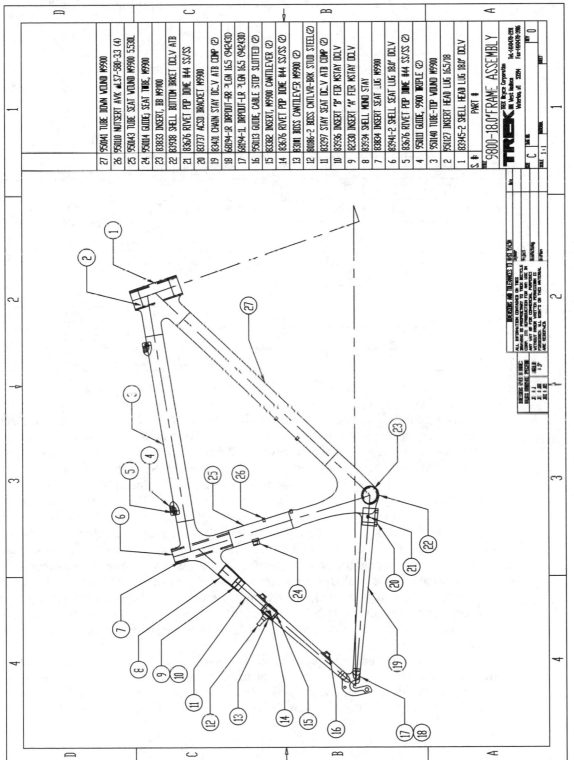

27	950041	TUBE DOWN VBOND M9900
26	950110	NUTSERT AVK ⌀L57-580-33 (4)
25	950043	TUBE SEAT VBOND M9900 5.530L
24	950114	GUIDE, SEAT TUBE, M9900
23	83833	INSERT, BB M9900
22	83938	SHELL BOTTOM BRKET DCLV ATB
21	83676	RIVET POP DOME #44 SS/SS
20	83737	ACSD BRACKET M9900
19	83401	CHAIN STAY DCLV ATB COMP (2)
18	68194-1R	DRPOUT-RR RLGN 16.5 (942430)
17	68194-1L	DRPOUT-LR RLGN 16.5 (942430)
16	950113	GUIDE, CABLE STOP SLOTTED (2)
15	83382	INSERT, M9900 CANTILEVER (2)
14	83676	RIVET POP DOME #44 SS/SS (2)
13	83101	BOSS CANTILEVER M9900
12	80186-2	BOSS CNTLVR-BRK STUD STEEL(2)
11	83397	STAY SEAT DCLV ATB COMP (2)
10	83936	INSERT "B" FER MSTAY DCLV
9	82330	INSERT "A" FER MSTAY DCLV
8	83934	SHELL MONO STAY
7	83834	INSERT SEAT JUG M9900
6	83941-2	SHELL SEAT LUG 18.0" DCLV
5	83676	RIVET POP DOME #44 SS/SS (2)
4	950111	GUIDE, 9900 TRIPLE (2)
3	950140	TUBE-TOP VBOND M9900
2	950127	INSERT HEAD LUG 16.5/18
1	83945-2	SHELL HEAD LUG 18.0" DCLV
S.#	PART #	

9800-18.0"FRAME ASSEMBLY

Figure 10-18 Courtesy of TREK Bicycle Corp.

Figure 10-19 shows another example of an assembly drawing parts list. The information in this parts list is:

- the detail number
- the part name or description
- the quantity of each part required
- the part material
- the part number (or drawing number)

NO	PART NAME	REQ'D	MATERIAL	DRAWING #
1	HOUSING RIGHT HAND	1	PLASTIC	A102994
2	HOUSING LEFT HAND	1	PLASTIC	A102894
3	SHAFT	1	STEEL	A91594
4	PINION	1	STD	NA
5	GEAR	1	STD	NA
6	BEARING	2	STD	NA
7	BEARING CUP	2	STD	NA
8	WASHER	2	STD	NA
9	PIN	2	STD	NA

Figure 10-19

An assembly drawing is often drawn in full or half section in order to better depict internal components and features. If the assembly is a multi-view drawing, then other section views such as removed, revolved, and broken-out may be utilized as required. Section 8.3 discussed the conventions used when assemblies are shown in section.

Another common type of assembly drawing is the pictorial assembly. Pictorial assemblies are typically drawn in isometric orientation. They may be shown in an "exploded" or "unexploded" format. An exploded pictorial shows the system with its parts separated from one another. This is done to show graphically the orientation and order in which the parts are assembled. Figure 10-20 shows an exploded pictorial assembly of a multi-positional tool vise. Note that the exploded positions of the components is usually along the axis of aligning features such as the centerlines of holes.

If the assembly is created from computer-generated solid model files, the pictorial assembly may show the positioning of the part models with respect to one another. This solid model may also be rendered to provide a realistic looking image of the assembled system.

Both exploded and unexploded pictorial assembly drawings may be used in service manuals, end-use customer assembly instructions, and product promotion materials. As with conventional pictorial views, hidden lines and centerlines (except for those used to depict part alignment) are often omitted from pictorial isometric assemblies.

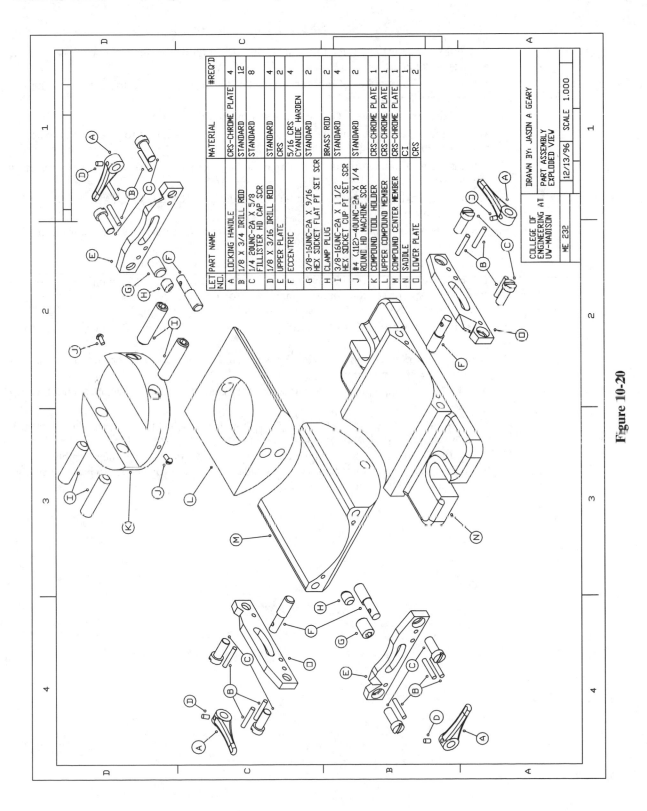

LET NO.	PART NAME	MATERIAL	#REQ'D
A	LOCKING HANDLE	CRS-CHROME PLATE	4
B	1/8 X 3/4 DRILL ROD	STANDARD	12
C	1/4 20UNC-2A X 5/8 FILLISTER HD CAP SCR	STANDARD	8
D	1/8 X 3/16 DRILL ROD	STANDARD	4
E	UPPER PLATE	CRS	2
F	ECCENTRIC	5/16 CRS CYANIDE HARDEN	4
G	3/8-15UNC-2A X 9/16 HEX SOCKET FLAT PT SET SCR	STANDARD	2
H	CLAMP PLUG	BRASS ROD	2
I	3/8-16UNC-2A X 1 1/2 HEX SOCKET CUP PT SET SCR	STANDARD	4
J	#4 (.112)-40UNC-2A X 1/4 ROUND HD MACHINE SCR	STANDARD	2
K	COMPOUND TOOL HOLDER	CRS-CHROME PLATE	1
L	UPPER COMPOUND MEMBER	CRS-CHROME PLATE	1
M	COMPOUND CENTER MEMBER	CRS-CHROME PLATE	1
N	SADDLE	CI	1
O	LOWER PLATE	CRS	2

COLLEGE OF ENGINEERING AT UW-MADISON	DRAWN BY: JASON A GEARY
	PART ASSEMBLY EXPLODED VIEW
ME 232	12/13/96	SCALE 1.000

Figure 10-20

EXERCISES 10.2

The following questions refer to the assembly drawing of the electric motor base and the associated detail drawings in Figures 10-21 and 10-22.

1. What is the total number of different parts that would be required to make one complete motor base assembly?

2. Which parts are purchased? Give your answer by the detail numbers.

3. With a freehand sketch, show the approximate procedure required to assemble the Support Shaft (#2) into the Base (#1).

4. What is the maximum horizontal travel that the Support Shaft can slide (horizontal travel) in the Base?

5. What is the maximum width between bolt holes (center to center) which a motor attached to this base can have?

6. What is the maximum length between bolt holes (center to center) which a motor attached to this base can have?

7. What size (diameter) fastener should be used in attaching a motor to the base?

8. The two given views of the Motor Bracket (#5) do not completely describe the shape of that part. What type of section could be employed to complete the shape description with the LEAST AMOUNT of alteration or changing of the current drawing?

9. What is the total horizontal adjustment possible between the Motor Bracket and the Motor Support?

10. With the Motor Support in a horizontal position, how far is it from the bottom of the Base to the top of the Motor Support?

11. Are the Motor Supports free to rotate about the Guide Shaft? If yes, what function(s) does this motion serve?

12. With a motor bolted in place, the two Motor Supports are fixed in position with respect to one another. What part, or parts, if any prevent sliding along the Guide Shaft?

13. Would the dimensions as given on the Base and the Guide Shaft permit the assembly of these two parts? If not, which dimensions on the Guide Shaft would you change to permit assembly and not affect the function of the part?

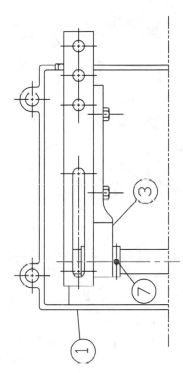

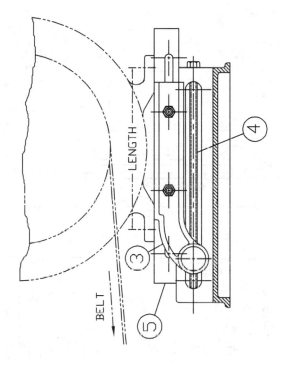

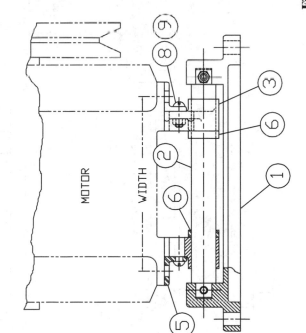

AS21789	MANSOR ELECTRIC CO. HOUGHTON MICHIGAN		SCALE: FULL
	DR. BY: F. SMITH		6/30/'90
	CHK BY: A. L. WARNER		
NO	NAME	MAT'L	REQ'D
1	BASE	CI	1
2	GUIDE SHAFT	STEEL	1
3	MOTOR SUPPORT (LH &RH)	CI	2
4	ADJUSTING SCREW	STEEL	2
5	MOTOR BRACKET (LH &RH)	STEEL	2
6	COLLAR	STEEL	2
7	.250x.250-20UNC-2A, SLOT HD, SET SCR	STD	2
8	.375-20UNC-2B, HEX NUT	STD	4
9	1.000x.375-16UNC-2A, RND HD, CAP SCR	STD	4

Figure 10-21

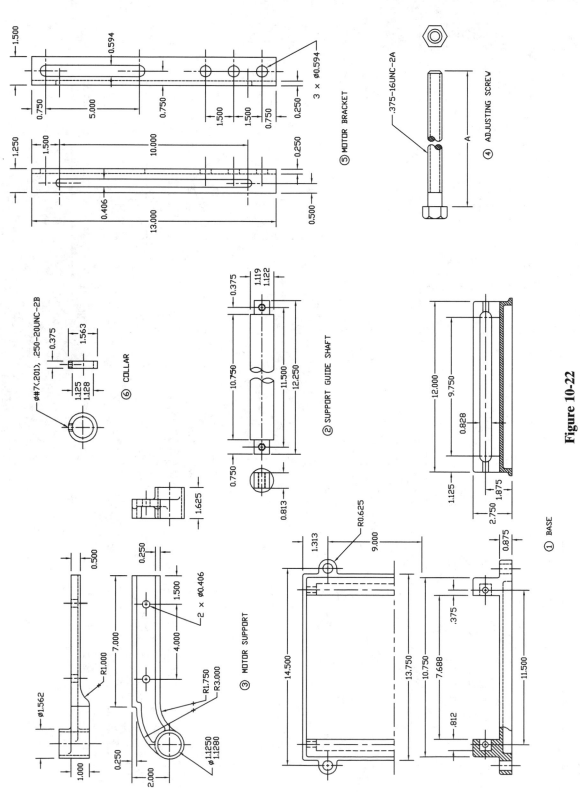

Figure 10-22

INDEX

- -

American National Standards Institute
(ANSI), 191–192
Angular dimensions, 194
Architect's scale, 121–122
Arrowheads, 193
Assembly drawing, 246
Assembly section, 185

Bearing, 15, 221, 230
Bench mark, 221–222
Bilateral tolerance, 218
Blueprints, 4, 220

Cabinet pictorial, 34
Cavalier pictorial, 34
Centerlines, 108
Conventional practice, 125–127
Coordinate planes, 13
Crosshatching, 166–167
Cut and fill, 187
Cutting plane line, 168
Cutting plane, 166
Cylindrical boundary, 108
Cylindrical intersections, 127–128

Decimal dimensions, 200
Decimal precision, 200
Departures, 16–17
Design layout, 240–241
Detail drawings, 241–242

Detail section, 228
Diametral dimensions, 195
Dilation factor, 48
Dilation, 48
Dimension lines, 192–193

Edge, 28, 72, 74
Edge-view, 142, 155
Elevation views, 220, 222
Engineer's scale, 119–121
Equations for planes, 14
Extension lines, 192–193

Floor plan, 220, 226
Foundation plan, 220, 222, 225–226
Four-view drawing, 138

General pictorial, 34
Geometric breakdown, 209–211
Glass cube, 71, 112, 132, 134–135

Hidden lines, 74–77
Hole patterns, 217
Horizontal view, 144–146

Inclined surfaces, 88–91
Isometric grid paper, 8
Isometric coordinate axis, 6–7
Isometric view, 27–30
    from orthographic views, 79

Latitudes, 16–17
Leader line, 197–199
Lettering technique, 38
Line drawing, 3
Linear dimension, 192
Lines, 146
    True length line, 147, 149–151
    Horizontal line, 147
    implied intersection, 12
    Profile line, 147
    Frontal line, 147
Local note (also callout), 197–199
Location dimensions, 211

Mating dimensions, 215–216
Metric scale, 112–123

Normal surface, 73–74

Oblique coordinate axis, 6–7
Oblique pictorial, 33–35
Oblique surfaces, 98–99
One-view drawing, 131
Orthographic views, 70–72
    plan view, 70, 220
    front elevation, 70
    side elevation, 70

Parts list, 246, 248
Pattern development, 64–66
Pictorial assembly, 248
Planes, 151
    True size, 153
    Horizontal, 153
    Frontal, 153
    Profile, 153
Point view, 148, 150
Principal views, 72
Profile view, 144–146
Projector (also projection rays), 71, 136

Radial dimensions, 196
Reference dimension, 218
Reflection, 60–62
Rib features, 183–184
Right-hand rule, 7
Rotation, 49–51, 54–56

Scale, 117–119
Schematic drawing, 3, 39–41
Section views
    Full, 169
    Half, 170
    Broken-out, 171
    Removed, 172
    Revolved, 173
    Offset, 173
    Aligned, 174–176
Single-curve surfaces, 108–110, 111–112
Site plan, 220, 222
Size dimensions, 211
Sketching
    lines, 25–26
    circles, 26
    ellipses, 26, 110
Stretch-out line, 66

Tabular drawing, 242
Titleblock, 240
Tolerance, 214–216, 218
Translation, 44–47
Traverse, 15, 17
True-size of surface, 135, 138–141, 155
Two-view drawing, 130

Unidirectional dimension, 99–200
Unilateral tolerance, 218
United States Geological Survey (USGS), 221

Wireframe geometry, 20–21